CONFORMAL INVARIANTS

TOPICS IN GEOMETRIC FUNCTION THEORY

CONFORMAL INVARIANTS

TOPICS IN GEOMETRIC FUNCTION THEORY

LARS V. AHLFORS

AMS CHELSEA PUBLISHING

American Mathematical Society • Providence, Rhode Island

2000 *Mathematics Subject Classification.* Primary 30–02.

For additional information and updates on this book, visit
www.ams.org/bookpages/chel-371

Library of Congress Cataloging-in-Publication Data

Ahlfors, Lars V. (Lars Valerian), 1907–1996.
    Conformal invariants : topics in geometric function theory / Lars V. Ahlfors.
        p. cm.
    Originally published: New York : McGraw-Hill, 1973, in series: McGraw-Hill series in higher mathematics.
    Includes bibliographical references and index.
    ISBN 978-0-8218-5270-5 (alk. paper)
    1. Conformal invariants.   2. Functions of complex variables.   3. Geometric function theory.
4. Riemann surfaces.   I. Title.

QA331.A46   2010
515′.9—dc22

                                                                                                2010035576

Contents

Foreword   ix

Preface   xi

1. Applications of Schwarz's lemma   1

  1-1. The noneuclidean metric   1

  1-2. The Schwarz-Pick theorem   3

  1-3. Convex regions   5

  1-4. Angular derivatives   7

  1-5. Ultrahyperbolic metrics   12

  1-6. Bloch's theorem   14

  1-7. The Poincaré metric of a region   16

  1-8. An elementary lower bound   16

  1-9. The Picard theorems   19

2. Capacity   23

  2-1. The transfinite diameter   23

  2-2. Potentials   24

  2-3. Capacity and the transfinite diameter   27

  2-4. Subsets of a circle   30

  2-5. Symmetrization   31

3. Harmonic measure   37

  3-1. The majorization principle   37

  3-2. Applications in a half plane   40

  3-3. Milloux's problem   41

  3-4. The precise form of Hadamard's theorem   44

4. Extremal length   50

  4-1. Definition of extremal length   50

  4-2. Examples   52

  4-3. The comparison principle   53

4-4.	The composition laws	54
4-5.	An integral inequality	56
4-6.	Prime ends	57
4-7.	Extremal metrics	61
4-8.	A case of spherical extremal metric	63
4-9.	The explicit formula for extremal distance	65
4-10.	Configurations with a single modulus	70
4-11.	Extremal annuli	71
4-12.	The function $\Lambda(R)$	74
4-13.	A distortion theorem	76
4-14.	Reduced extremal distance	78

5. Elementary theory of univalent functions **82**

5-1.	The area theorem	82
5-2.	The Grunsky and Golusin inequalities	85
5-3.	Proof of $|a_4| \leq 4$	87

6. Löewner's method **92**

6-1.	Approximation by slit mappings	92
6-2.	Löewner's differential equation	96
6-3.	Proof of $|a_3| \leq 3$	96

7. The Schiffer variation **98**

7-1.	Variation of the Green's function	98
7-2.	Variation of the mapping function	102
7-3.	The final theorem	105
7-4.	The slit variation	106

8. Properties of the extremal functions **107**

8-1.	The differential equation	107
8-2.	Trajectories	110
8-3.	The Γ structures	114

8-4.	Regularity and global correspondence	116
8-5.	The case $n = 3$	118

9. Riemann surfaces — **125**

9-1.	Definition and examples	125
9-2.	Covering surfaces	127
9-3.	The fundamental group	128
9-4.	Subgroups and covering surfaces	130
9-5.	Cover transformations	132
9-6.	Simply connected surfaces	134

10. The uniformization theorem — **136**

10-1.	Existence of the Green's function	136
10-2.	Harmonic measure and the maximum principle	138
10-3.	Equivalence of the basic conditions	139
10-4.	Proof of the uniformization theorem (Part I)	142
10-5.	Proof of the uniformization theorem (Part II)	147
10-6.	Arbitrary Riemann surfaces	149
	Bibliography	152
	Index	156
	Errata	159

Foreword

Lars Ahlfors often spoke of his excitement as a young student listening to Rolf Nevanlinna's lectures on the new theory of meromorphic functions. It was, as he writes in his collected papers, his "first exposure to live mathematics." In his enormously influential research papers and in his equally influential books, Ahlfors shared with the reader, both professional and student, that excitement.

The present volume derives from lectures given at Harvard over many years, and the topics would now be considered quite classical. At the time the book was published, in 1973, most of the results were already decades old. Nevertheless, the mathematics feels very much alive and still exciting, for one hears clearly the voice of a master speaking with deep understanding of the importance of the ideas that make up the course.

Moreover, several of those ideas originated with or were cultivated by the author. The opening chapter on Schwarz's lemma contains Ahlfors' celebrated discovery, from 1938, of the connection between that very classical result and conformal metrics of negative curvature. The theme of using conformal metrics in connection with conformal mapping is elucidated in the longest chapter of the book, on extremal length. It would be hard to overstate the impact of that method, but until the book's publication there were very few places to find a coherent exposition of the main ideas and applications. Ahlfors credited Arne Beurling as the principal originator, and with the publication of Beurling's collected papers [2] one now has access to some of his own reflections.

Extremal problems are a recurring theme, and this strongly influences the choices Ahlfors makes throughout the book. Capacity is often discussed in relation to small point sets in function theory, with implications for existence theorems, but in that chapter Ahlfors has a different goal, aiming instead for the solution of a geometric extremal problem on closed subsets of the unit circle. The method of harmonic measure appeals to the Euclidean geometry of a domain and parts of its boundary to systematize the use of the maximum principle. Here Ahlfors concentrates on two problems, Milloux's problem, as treated in Beurling's landmark thesis, and a precise version of Hadamard's three circles theorem in a form given by Teichmüller. Nowhere else is there an accessible version of Teichmüller's solution. The chapter on harmonic measure provides only a small sample of a large circle of ideas, developed more systematically in the recent book [7].

Ahlfors devotes four short chapters to discussions of extremal problems for univalent functions, with focus on Loewner's parametric method and Schiffer's variational method. The material on coefficient estimates is now quite dated, following the proof of the Bieberbach conjecture by Louis de Branges [3] and its subsequent adaptation [6] appealing to the classical form of Loewner's differential equation. However, the methods of Loewner and Schiffer have broad applications in geometric function theory and their relevance is undiminished. More detailed treatments have since appeared [8,4], but Ahlfors' overview still brings these ideas to life. In recent years, Loewner's method has stepped into the limelight again with Oded Schramm's discovery of the stochastic Loewner equation and its connections with mathematical physics.

The final two chapters give an introduction to Riemann surfaces, with topological and analytical background supplied to support a proof of the uniformization theorem. In the author's treatment, as in all treatments, the main difficulty is in the parabolic case. Overall, the reader is encouraged to consult other sources for more details, for example [5].

We close with Ahlfors' own words from an address in 1953 at a conference celebrating the centennial of Riemann's dissertation [1]:

> Geometric function theory of one variable is already a highly developed branch of mathematics, and it is not one in which an easily formulated classical problem awaits its solution. On the contrary it is a field in which the formulation of essential problems is almost as important as their solution; it is a subject in which methods and principles are all-important, while an isolated result, however pretty and however difficult to prove, carries little weight.

The reader can learn much of this from the present volume. Furthermore, Ahlfors' remarks came around the time that quasiconformal mappings and, later, Kleinian groups began to flower, fields in which he was the leader. What a second volume those topics would have made!

Peter Duren

F. W. Gehring

Brad Osgood

References

[1] *Lars Valerian Ahlfors: Collected Papers*, Birkhäuser, Boston, 1982.

[2] *Collected Works of Arne Beurling*, edited by L. Carleson, P. Malliavin, J. Neuberger, and J. Wermer, Birkhäuser, Boston, 1989.

[3] L. de Branges, A proof of the Bieberbach conjecture, *Acta Math.* **154** (1985), 137–152.

[4] P. L. Duren, *Univalent Functions*, Springer-Verlag, New York, 1983.

[5] H. Farkas and I. Kra, *Riemann Surfaces*, Second Edition, Springer-Verlag, New York, 1991.

[6] C. H. FitzGerald and Ch. Pommerenke, The de Branges theorem on univalent functions, *Trans. Amer. Math. Soc.* **290** (1985), 683–690.

[7] J. B. Garnett and D. E. Marshall, *Harmonic Measure*, Cambridge University Press, New York, 2005.

[8] Ch. Pommerenke, *Univalent Functions*, Vandenhoeck & Ruprecht, Göttingen, 1975.

Preface

This is a textbook primarily intended for students with approximately a year's background in complex variable theory. The material has been collected from lecture courses given over a long period of years, mostly at Harvard University. The book emphasizes classic and semiclassic results which the author feels every student of complex analysis should know before embarking on independent research. The selection of topics is rather arbitrary, but reflects the author's preference for the geometric approach. There is no attempt to cover recent advances in more specialized directions.

Most conformal invariants can be described in terms of extremal properties. Conformal invariants and extremal problems are therefore intimately linked and form together the central theme of this book. An obvious reason for publishing these lectures is the fact that much of the material has never appeared in textbook form. In particular this is true of the theory of extremal length, instigated by Arne Beurling, which should really be the subject of a monograph of its own, preferably by Beurling himself. Another topic that has received only scant attention in the textbook literature is Schiffer's variational method, which I have tried to cover as carefully and as thoroughly as I know how. I hope very much that this account will prove readable. I have also included a new proof of $|a_4| \leq 4$ which appeared earlier in a *Festschrift* for M. A. Lavrentiev (in Russian).

The last two chapters, on Riemann surfaces, stand somewhat apart from the rest of the book. They are motivated by the need for a quicker approach to the uniformization theorem than can be obtained from Leo Sario's and my book "Riemann Surfaces."

Some early lectures of mine at Oklahoma A. and M. College had been transcribed by R. Osserman and M. Gerstenhaber, as was a lecture at Harvard University on extremal methods by E. Schlesinger. These writeups were of great help in assembling the present version. I also express my gratitude to F. Gehring without whose encouragement I would not have gone ahead with publication.

There is some overlap with Makoto Ohtsuka's book "Dirichlet Problem, Extremal Length and Prime Ends" (Van Nostrand, 1970) which is partly based on my lectures at Harvard University and in Japan.

Lars V. Ahlfors

1

APPLICATIONS OF SCHWARZ'S LEMMA

1-1 THE NONEUCLIDEAN METRIC

The fractional linear transformation

$$S(z) = \frac{az + b}{\bar{b}z + \bar{a}} \tag{1-1}$$

with $|a|^2 - |b|^2 = 1$ maps the unit disk $\Delta = \{z; |z| < 1\}$ conformally onto itself. It is also customary to write (1-1) in the form

$$S(z) = e^{i\alpha} \frac{z - z_0}{1 - \bar{z}_0 z} \tag{1-2}$$

which has the advantage of exhibiting $z_0 = S^{-1}(0)$ and $\alpha = \arg S'(0)$.

Consider $z_1, z_2 \in \Delta$ and set $w_1 = S(z_1)$, $w_2 = S(z_2)$. From (1-1) we obtain

$$w_1 - w_2 = \frac{z_1 - z_2}{(\bar{b}z_1 + \bar{a})(\bar{b}z_2 + \bar{a})}$$

$$1 - \bar{w}_1 w_2 = \frac{1 - \bar{z}_1 z_2}{(b\bar{z}_1 + a)(\bar{b}z_2 + \bar{a})},$$

and hence
$$\left| \frac{z_1 - z_2}{1 - \bar{z}_1 z_2} \right| = \left| \frac{w_1 - w_2}{1 - \bar{w}_1 w_2} \right|. \qquad (1\text{-}3)$$

We say that
$$\delta(z_1, z_2) = \left| \frac{z_1 - z_2}{1 - \bar{z}_1 z_2} \right| \qquad (1\text{-}4)$$

is a *conformal invariant*. Comparison of (1-2) and (1-4) shows that $\delta(z_1, z_2) < 1$, a fact that can also be read off from the useful identity

$$1 - \delta(z_1, z_2)^2 = \frac{(1 - |z_1|^2)(1 - |z_2|^2)}{|1 - \bar{z}_1 z_2|^2}.$$

If z_1 approaches z_2, (1-3) becomes

$$\frac{|dz|}{1 - |z|^2} = \frac{|dw|}{1 - |w|^2}.$$

This shows that the Riemannian metric whose element of length is

$$ds = \frac{2|dz|}{1 - |z|^2} \qquad (1\text{-}5)$$

is invariant under conformal self-mappings of the disk (the reason for the factor 2 will become apparent later). In this metric every rectifiable arc γ has an invariant length

$$\int_\gamma \frac{2|dz|}{1 - |z|^2},$$

and every measurable set E has an invariant area

$$\iint_E \frac{4dx\,dy}{(1 - |z|^2)^2}.$$

The shortest arc from 0 to any other point is along a radius. Hence the geodesics are circles orthogonal to $|z| = 1$. They can be considered straight lines in a geometry, the *hyperbolic* or *noneuclidean* geometry of the disk.

The noneuclidean distance from 0 to $r > 0$ is

$$\int_0^r \frac{2dr}{1 - r^2} = \log \frac{1 + r}{1 - r}.$$

Since $\delta(0, r) = r$, it follows that the noneuclidean distance $d(z_1, z_2)$ is connected with $\delta(z_1, z_2)$ through $\delta = \tanh (d/2)$.

The noneuclidean geometry can also be carried over to the half plane

$H = \{z = x + iy; y > 0\}$. The element of length that corresponds to the choice (1-5) is

$$ds = \frac{|dz|}{y},\qquad(1\text{-}6)$$

and the straight lines are circles and lines orthogonal to the real axis.

1-2 THE SCHWARZ–PICK THEOREM

The classic Schwarz lemma asserts the following: If f is analytic and $|f(z)| < 1$ for $|z| < 1$, and if $f(0) = 0$, then $|f(z)| \leq |z|$ and $|f'(0)| \leq 1$. Equality $|f(z)| = |z|$ with $z \neq 0$ or $|f'(0)| = 1$ can occur only for $f(z) = e^{i\alpha}z$, α a real constant.

There is no need to reproduce the well-known proof. It was noted by Pick that the result can be expressed in invariant form.

Theorem 1-1 An analytic mapping of the unit disk into itself decreases the noneuclidean distance between two points, the noneuclidean length of an arc, and the noneuclidean area of a set.

The explicit inequalities are

$$\frac{|f(z_1) - f(z_2)|}{|1 - \overline{f(z_1)}f(z_2)|} \leq \frac{|z_1 - z_2|}{|1 - \bar{z}_1 z_2|}$$

$$\frac{|f'(z)|}{1 - |f(z)|^2} \leq \frac{1}{1 - |z|^2}.$$

Nontrivial equality holds only when f is a fractional linear transformation of the form (1-1).

Pick does not stop with this observation. He also proves the following more general version which deserves to be better known.

Theorem 1-2 Let $f: \Delta \to \Delta$ be analytic and set $w_k = f(z_k)$, $k = 1, \ldots, n$. Then the Hermitian form

$$Q_n(t) = \sum_{h,k=1}^{n} \frac{1 - w_h\bar{w}_k}{1 - z_h\bar{z}_k}\, t_h\bar{t}_k$$

is positive definite (or semidefinite).

PROOF We assume first that f is analytic on the closed disk. The function $F = (1 + f)/(1 - f)$ has a positive real part, and if $F = U + iV$

we have the representation

$$F(z) = \frac{1}{2\pi} \int_0^{2\pi} \frac{e^{i\theta} + z}{e^{i\theta} - z} \, U(e^{i\theta}) \, d\theta + iV(0).$$

This gives

$$F(z_h) + \overline{F(z_k)} = \frac{1}{\pi} \int_0^{2\pi} \frac{1 - z_h \bar{z}_k}{(e^{i\theta} - z_h)(e^{-i\theta} - \bar{z}_k)} \, U \, d\theta,$$

and hence

$$\sum_{h,k=1}^n \frac{F_h + \bar{F}_k}{1 - z_h \bar{z}_k} t_h \bar{t}_k = \frac{1}{\pi} \int_0^{2\pi} \left| \sum_1^n \frac{t_k}{e^{i\theta} - z_k} \right|^2 U \, d\theta \geq 0.$$

Here $F_h + \bar{F}_k = 2(1 - w_h \bar{w}_k)/(1 - w_h)(1 - \bar{w}_k)$. The factors in the denominator can be incorporated in t_h, $\bar{t}_k$, and we conclude that $Q_n(t) \geq 0$. For arbitrary f we apply the theorem to $f(rz)$, $0 < r < 1$, and pass to the limit.

Explicitly, the condition means that all the determinants

$$D_k = \begin{vmatrix} \dfrac{1 - |w_1|^2}{1 - |z_1|^2} & \cdots & \dfrac{1 - w_1 \bar{w}_k}{1 - z_1 \bar{z}_k} \\ \cdots & \cdots & \cdots \\ \dfrac{1 - w_k \bar{w}_1}{1 - z_k \bar{z}_1} & \cdots & \dfrac{1 - |w_k|^2}{1 - |z_k|^2} \end{vmatrix}$$

are ≥ 0. It can be shown that these conditions are also sufficient for the interpolation problem to have a solution. If $w_1, \ldots, w_{n-1}$ are given and $D_1, \ldots, D_{n-1} \geq 0$, the condition on w_n will be of the form $|w_n|^2 + 2 \operatorname{Re}(a w_n) + b \leq 0$. This means that w_n is restricted to a certain closed disk. It turns out that the disk reduces to a point if and only if $D_{n-1} = 0$.

The proof of the sufficiency is somewhat complicated and would lead too far from our central theme. We shall be content to show, by a method due to R. Nevanlinna, that the possible values of w_n fill a closed disk. We do not prove that this disk is determined by $D_n \geq 0$.

Nevanlinna's reasoning is recursive. For $n = 1$ there is very little to prove. Indeed, there is no solution if $|w_1| > 1$. If $|w_1| = 1$ there is a unique solution, namely, the constant w_1. If $|w_1| < 1$ and f_1 is a solution, then

$$f_2(z) = \frac{f_1(z) - w_1}{1 - \bar{w}_1 f_1(z)} : \frac{z - z_1}{1 - \bar{z}_1 z} \tag{1-7}$$

is regular in Δ, and we have proved that $|f_2(z)| \leq 1$. Conversely, for any such function f_2 formula (1-7) yields a solution f_1.

For $n = 2$ the solutions, if any, are among the functions f_1 already determined, and $f_2(z_2)$ must be equal to a prescribed value $w_2^{(2)}$. There are the same alternatives as before, and it is clear how the process continues. We are trying to construct a sequence of functions f_k of modulus ≤ 1 with certain prescribed values $f_k(z_k) = w_k^{(k)}$ which can be calculated from $w_1, \ldots, w_k$. If $|w_k^{(k)}| > 1$ for some k, the process comes to a halt and there is no solution. If $|w_k^{(k)}| = 1$, there is a unique f_k, and hence a unique solution of the interpolation problem restricted to $z_1, \ldots, z_k$. In case all $|w_k^{(k)}| < 1$, the recursive relations

$$f_{k+1}(z) = \frac{f_k(z) - w_k^{(k)}}{1 - \bar{w}_k^{(k)} f_k(z)} : \frac{z - z_k}{1 - \bar{z}_k z} \qquad k = 1, \ldots, n$$

lead to all solutions f_1 of the original problem when f_{n+1} ranges over all analytic functions with $|f_{n+1}(z)| \leq 1$ in Δ.

Because the connection between f_k and f_{k+1} is given as a fractional linear transformation, the general solution is of the form

$$f_1(z) = \frac{A_n(z) f_{n+1}(z) + B_n(z)}{C_n(z) f_{n+1}(z) + D_n(z)},$$

where A_n, B_n, C_n, D_n are polynomials of degree n determined by the data of the problem. We recognize now that the possible values of $f(z)$ at a fixed point do indeed range over a closed disk.

This solution was given in R. Nevanlinna [42]. The corresponding problem for infinitely many z_k, w_k was studied by Denjoy [17], R. Nevanlinna [43], and more recently Carleson [13].

1-3   CONVEX REGIONS

A set is convex if it contains the line segment between any two of its points. We wish to characterize the analytic functions f that define a one-to-one conformal map of the unit disk on a convex region. For simplicity such functions will be called convex univalent (Hayman [27]).

Theorem 1-3   An analytic function f in Δ is convex univalent if and only if

$$\mathrm{Re}\, \frac{zf''(z)}{f'(z)} \geq -1 \qquad\qquad (1\text{-}8)$$

for all $z \in \Delta$. When this is true the stronger inequality

$$\left| \frac{zf''(z)}{f'(z)} - \frac{2|z|^2}{1 - |z|^2} \right| \leq \frac{2|z|}{1 - |z|^2} \qquad\qquad (1\text{-}9)$$

is also in force.

Suppose for a moment that f is not only convex univalent but also analytic on the closed disk. It is intuitively clear that the image of the unit circle has a tangent which turns in the positive direction when $\theta = \arg z$ increases. This condition is expressed through $\partial/\partial\theta \arg df \geq 0$. But $\arg df = \arg f' + \arg dz = \arg f' + \theta + \pi/2$, and the condition becomes $\partial/\partial\theta (\arg f' + \theta) = \mathrm{Re}\, (zf''/f' + 1) \geq 0$ for $|z| = 1$. By the maximum principle the same holds for $|z| < 1$.

Although this could be made into a rigorous proof, we much prefer an idea due to Hayman. We may assume that $f(0) = 0$. If f is convex univalent, the function

$$g(z) = f^{-1}\left[\frac{f(\sqrt{z}) + f(-\sqrt{z})}{2}\right]$$

is well defined, analytic, and of absolute value <1 in Δ. Hence $|g'(0)| \leq 1$. But if $f(z) = a_1 z + a_2 z^2 + \cdots$, then $g(z) = (a_2/a_1)z + \cdots$, and we obtain $|a_2/a_1| \leq 1$, $|f''(0)/f'(0)| \leq 2$. This is (1-9) for $z = 0$.

We apply this result to $F(z) = f[(z + c)/(1 + \bar{c}z)]$, $|c| < 1$, which maps Δ on the same region. Simple calculations give

$$\frac{F''(0)}{F'(0)} = \frac{f''(c)}{f'(c)}\,(1 - |c|^2) - 2\bar{c},$$

and we obtain (1-9) and its consequence (1-8).

The proof of the converse is less elegant. It is evidently sufficient to prove that the image of $\Delta_r = \{z; |z| < r\}$ is convex for every $r < 1$. The assumption (1-8) implies that $\arg df$ increases with θ on $|z| = r$. Since f' is never zero, the change of $\arg df$ is 2π. Therefore, we can find θ_1 and θ_2 such that $\arg df$ increases from 0 to π on $[\theta_1, \theta_2]$ and from π to 2π on $[\theta_2, \theta_1 + 2\pi]$. If $f(re^{i\theta}) = u(\theta) + iv(\theta)$, it follows that v increases on the first interval and decreases on the second. Let v_0 be a real number between the minimum $v(\theta_1)$ and the maximum $v(\theta_2)$. Then $v(\theta)$ passes through v_0 exactly once on each of the intervals, and routine use of winding numbers shows that the image of Δ_r intersects the line $v = v_0$ along a single segment. The same reasoning applies to parallels in any direction, and we conclude that the image is convex.

The condition $|f''(0)/f'(0)| \leq 2$ has an interesting geometric interpretation. Consider an arc γ in Δ that passes through the origin and whose image is a straight line. The curvature of γ is measured by $d(\arg dz)/|dz|$. By assumption $d(\arg df) = 0$ along γ so that $d(\arg dz) = -d\arg f'$. The curvature is thus a directional derivative of $\arg f'$, and as such it is at most $|f''/f'|$ in absolute value. We conclude that the curvature at the origin is at most 2.

This result has an invariant formulation. If the curvature at the origin is ≤ 2, the circle of curvature intersects $|z| = 1$. But the circle of curvature is the circle of highest contact. A conformal self-mapping preserves circles and preserves order of contact. Circles of curvature are mapped on circles of curvature, and our result holds not only at the origin, but at any point.

Theorem 1-4   Let γ be a curve in Δ whose image under a conformal mapping on a convex region is a straight line. Then the circles of curvature of γ meet $|z| = 1$.

This beautiful result is due to Carathéodory.

1-4   ANGULAR DERIVATIVES

For $|a| < 1$ and $R < 1$ let $K(a,R)$ be the set of all z such that

$$\left| \frac{z - a}{1 - \bar{a}z} \right| < R.$$

Clearly, $K(a,R)$ is an open noneuclidean disk with center a and radius d such that $R = \tanh(d/2)$.

Let $K_n = K(z_n,R_n)$ be a sequence of disks such that $z_n \to 1$ and

$$\frac{1 - |z_n|}{1 - R_n} \to k \neq 0, \infty. \tag{1-10}$$

We claim that the K_n tend to the *horocycle* K_∞ defined by

$$\frac{|1 - z|^2}{1 - |z|^2} < k. \tag{1-11}$$

The horocycle is a disk tangent to the unit circle at $z = 1$.

The statement $K_n \to K_\infty$ is to be understood in the following sense: (1) If $z \in K_n$ for infinitely many n, then $z \in \bar{K}_\infty$, the closure of K_∞; (2) if $z \in K_\infty$, then $z \in K_n$ for all sufficiently large n. For the proof we observe that $z \in K_n$ is equivalent to

$$\frac{|1 - \bar{z}_n z|^2}{1 - |z|^2} < \frac{1 - |z_n|^2}{1 - R_n^2}. \tag{1-12}$$

If this is true for infinitely many n, we can go to the limit and obtain (1-11) by virtue of (1-10), except that equality may hold. Conversely, if

(1-11) holds, then

$$\lim_{n \to \infty} \frac{|1 - \bar{z}_n z|^2}{1 - |z|^2} < k$$

while

$$\lim_{n \to \infty} \frac{1 - |z_n|^2}{1 - R_n{}^2} = k,$$

so that (1-12) must hold for all sufficiently large n.

After these preliminaries, let f be analytic and $|f(z)| < 1$ in Δ. Suppose that $z_n \to 1$, $f(z_n) \to 1$, and

$$\frac{1 - |f(z_n)|}{1 - |z_n|} \to \alpha \neq \infty. \tag{1-13}$$

Given $k > 0$ we choose R_n so that $(1 - |z_n|)/(1 - R_n) = k$; this makes $0 < R_n < 1$ provided $1 - |z_n| < k$. With the same notation

$$K_n = K(z_n, R_n)$$

as above, we know by Schwarz's lemma that $f(K_n) \subset K_n' = K(w_n, R_n)$ where $w_n = f(z_n)$. The K_n converge to the horocycle K_∞ with parameter k as in (1-11), and because $(1 - |w_n|)/(1 - R_n) \to \alpha k$, the K_n' converge to K_∞' with parameter αk. If $z \in K_\infty$, it belongs to infinitely many K_n. Hence $f(z)$ belongs to infinitely many K_n' and consequently to $\bar{K}_\infty'$. In view of the continuity it follows that

$$\frac{|1 - z|^2}{1 - |z|^2} \le k \qquad \text{implies} \qquad \frac{|1 - f(z)|^2}{1 - |f(z)|^2} \le \alpha k.$$

This is known as *Julia's lemma*.

Since k is arbitrary, the same result may be expressed by

$$\frac{|1 - f(z)|^2}{1 - |f(z)|^2} \le \alpha \frac{|1 - z|^2}{1 - |z|^2},$$

or by

$$\beta = \sup \left[\frac{|1 - f(z)|^2}{1 - |f(z)|^2} : \frac{|1 - z|^2}{1 - |z|^2} \right] \le \alpha.$$

In particular, α is never 0, and if $\beta = \infty$, there is no finite α.

Let us now assume $\beta < \infty$ and take $z_n = x_n$ to be real. Then

$$|1 - w_n|^2 < \beta \frac{1 - x_n}{1 + x_n},$$

and the condition $w_n \to 1$ is automatically fulfilled. Furthermore,

$$\beta \ge \frac{|1 - w_n|^2}{1 - |w_n|^2} \frac{1 + x_n}{1 - x_n} \ge \frac{1 + x_n}{1 + |w_n|} \frac{|1 - w_n|}{1 - x_n} \ge \frac{1 + x_n}{1 + |w_n|} \frac{1 - |w_n|}{1 - x_n}$$

so that (1-13) implies $\alpha \leq \beta$. Hence $\alpha = \beta$ for arbitrary approach along the real axis, and we conclude that

$$\lim_{x \to 1} \frac{1 - |f(x)|}{1 - x} = \lim_{x \to 1} \frac{|1 - f(x)|}{1 - x} = \beta. \tag{1-14}$$

Since $\beta \neq 0, \infty$, the equality of these limits easily implies $\arg [1 - f(x)] \to 0$, and with this information (1-14) can be improved to

$$\lim_{x \to 1} \frac{1 - f(x)}{1 - x} = \beta. \tag{1-15}$$

We have proved (1-14) and (1-15) only if $\beta \neq \infty$. However, if $\beta = \infty$, we know that (1-13) can never hold with a finite α. Hence (1-14) is still true, and for $\beta = \infty$ (1-14) implies (1-15).

So far we have shown that the quotient $[1 - f(z)]/(1 - z)$ always has a radial limit. We shall complete this result by showing that the quotient tends to the same limit when $z \to 1$ subject to a condition $|1 - z| \leq M(1 - |z|)$. The condition means that z stays within an angle less than π, and the limit is referred to as an angular limit.

Theorem 1-5   Suppose that f is analytic and $|f(z)| < 1$ in Δ. Then the quotient

$$\frac{1 - f(z)}{1 - z}$$

always has an angular limit for $z \to 1$. This limit is equal to the least upper bound of

$$\frac{|1 - f(z)|^2}{1 - |f(z)|^2} : \frac{|1 - z|^2}{1 - |z|^2},$$

and hence either $+\infty$ or a positive real number. If it is finite, $f'(z)$ has the same angular limit.

PROOF   We have to show that β is an angular limit. If $\beta = \infty$, no new reasoning is needed, for we conclude as before that

$$\lim_{z \to 1} \frac{1 - |f(z)|}{1 - |z|} = \infty,$$

and when $|1 - z| \leq M(1 - |z|)$ this implies

$$\lim_{z \to 1} \frac{1 - f(z)}{1 - z} = \infty.$$

The case of a finite β can be reduced to the case $\beta = \infty$. The definition of β as a least upper bound implies

$$\operatorname{Re} \frac{1+z}{1-z} \le \beta \operatorname{Re} \frac{1+f(z)}{1-f(z)}.$$

Therefore we can write

$$\beta \frac{1+f}{1-f} - \frac{1+z}{1-z} = \frac{1+F}{1-F} \tag{1-16}$$

with $|F| < 1$. Because β cannot be replaced by a smaller number, it is clear that the function F must fall under the case $\beta = \infty$ so that $(1-z)/(1-F) \to 0$ in every angle. It then follows from (1-16) that $(1-f)/(1-z)$ has the angular limit β.

From (1-16) we have further

$$\beta f'(1-f)^{-2} - (1-z)^{-2} = F'(1-F)^{-2}.$$

We know by Schwarz's lemma that $|F'|/(1-|F|^2) \le 1/(1-|z|^2)$. With this estimate, together with $|1-z| \le M(1-|z|)$, we obtain

$$\left| \beta f'(z) \left[\frac{1-z}{1-f(z)} \right]^2 - 1 \right| \le 2M^2 \frac{1-|z|}{1-|F|} \to 0,$$

and from this we conclude that $f'(z) \to \beta$.

When $\beta \ne \infty$, it is called the angular derivative at 1. In this case the limit $f(1) = 1$ exists as an angular limit, and β is the angular limit of the difference quotient $[f(z) - f(1)]/(z - 1)$ as well as of $f'(z)$. The mapping by f is conformal at $z = 1$ provided we stay within an angle.

The theorem may be applied to $f_1(z) = e^{-i\delta}f(e^{-i\gamma}z)$ with any real γ and δ, but it is of no interest unless $f(z) \to e^{i\delta}$ as $z \to e^{i\gamma}$ along a radius. In that case the difference quotient $[f(z) - e^{i\delta}]/(z - e^{i\gamma})$ has a finite limit, and the mapping is conformal at $e^{i\gamma}$ if this limit is different from zero.

In many cases it is more convenient to use half planes. For instance, if $f = u + iv$ maps the right half plane into itself, we are able to conclude that

$$\lim_{z \to \infty} \frac{f(z)}{z} = \lim_{z \to \infty} \frac{u(z)}{x} = c = \inf \frac{u(z)}{x}, \tag{1-17}$$

the limits being restricted to $|\arg z| \le \pi/2 - \epsilon, \epsilon > 0$. Indeed, if the theorem is applied to $f_1 = (f - 1)/(f + 1)$ as a function of $z_1 = (z - 1)/(z + 1)$, we have $\beta = \sup x/u = 1/c$ and

$$\lim_{z_1 \to 1} \frac{1-z_1}{1-f_1} = \lim_{z \to \infty} \frac{1+f}{1+z} = c.$$

This easily implies (1-17). Note that c is finite and ≥ 0.

The proof of Theorem 1-5 that we have given is due to Carathéodory [10]. We have chosen this proof because of its clear indication that the theorem is in fact a limiting case of Schwarz's lemma. There is another proof, based on the Herglotz representation of an analytic function with positive real part, which is perhaps even simpler. We recall the Poisson-Schwarz representation used in the proof of Theorem 1-2. For positive U it can be rewritten in the form

$$F(z) = \int_0^{2\pi} \frac{e^{i\theta} + z}{e^{i\theta} - z} d\mu(\theta) + iC,$$

where μ denotes a finite positive measure on the unit circle. In this form, as observed by Herglotz, it is valid for arbitrary analytic functions with a positive real part.

Apply the formula to $F = (1 + f)/(1 - f)$, where $|f(z)| < 1$ in Δ. Let $c \geq 0$ denote $\mu(\{0\})$, i.e., the part of μ concentrated at the point 1, and denote the rest of the measure by μ_0 so that we can write

$$\frac{1 + f}{1 - f} = c \frac{1 + z}{1 - z} + \int_0^{2\pi} \frac{e^{i\theta} + z}{e^{i\theta} - z} d\mu_0(\theta) + iC. \tag{1-18}$$

For the real parts we thus have

$$\frac{1 - |f|^2}{|1 - f|^2} = c \frac{1 - |z|^2}{|1 - z|^2} + \int_0^{2\pi} \frac{1 - |z|^2}{|e^{i\theta} - z|^2} d\mu_0(\theta), \tag{1-19}$$

from which it is already clear that

$$\frac{1 - |f|^2}{|1 - f|^2} \geq c \frac{1 - |z|^2}{|1 - z|^2}.$$

We rewrite (1-19) as

$$\frac{1 - |f|^2}{|1 - f|^2} : \frac{1 - |z|^2}{|1 - z|^2} = c + I(z)$$

with

$$I(z) = \int_0^{2\pi} \frac{|1 - z|^2}{|e^{i\theta} - z|^2} d\mu_0(\theta).$$

We claim that $I(z) \to 0$ as $z \to 1$ in an angle. For this purpose we choose δ so small that the μ_0 measure of the interval $(-\delta, \delta)$ is less than a given $\epsilon > 0$. Divide $I(z)$ in two parts:

$$I = I_0 + I_1 = \int_{-\delta}^{\delta} + \int_{\delta}^{2\pi - \delta}.$$

If $|1 - z| \leq M(1 - |z|)$, it is immediate that $|I_0| \leq M^2\epsilon$. It is obvious that $I_1 \to 0$, and we conclude that $I(z) \to 0$ in an angle. This proves that $c = 1/\beta$ in the earlier notation

If the same reasoning is applied directly to (1-18), we find that $(1 - z)(1 + f)(1 - f)^{-1} \to 2c$ in an angle, and this is equivalent to $(1 - z)/(1 - f) \to c$. This completes the alternate proof of Theorem 1-5.

As an application we shall prove a theorem known as Löwner's lemma. As before, f will be an analytic mapping of Δ into itself, but this time we add the assumption that $|f(z)| \to 1$ as z approaches an open arc γ on $|z| = 1$. Then f has an analytic extension to γ by virtue of the reflection principle, and $f'(\zeta) \neq 0$ for $\zeta \in \gamma$. Indeed, if $f'(\zeta)$ were zero the value $f(\zeta)$ would be assumed with multiplicity greater than 1, and this is incompatible with $|f(\zeta)| = 1$ and $|f(z)| < 1$ for $|z| < 1$. It is also true that arg $f(\zeta)$ increases with arg ζ so that f defines a locally one-to-one mapping of γ on an arc γ'.

Theorem 1-6   If in these circumstances $f(0) = 0$, then the length of γ' is at least equal to the length of γ.

PROOF   We apply Theorem 1-5 to $F(z) = f(\zeta z)/f(\zeta)$, $\zeta \in \gamma$. The angular derivative at $z = 1$ is

$$\lim_{r \to 1} \frac{1 - F(r)}{1 - r} = F'(1) = \frac{\zeta f'(\zeta)}{f(\zeta)} = |f'(\zeta)|,$$

for arg $f'(\zeta) = $ arg $[f(\zeta)/\zeta]$. But $|1 - F(r)| \geq 1 - |F(r)| \geq 1 - r$ by Schwarz's lemma. Hence $|f'(\zeta)| \geq 1$, and the theorem follows.

1-5   ULTRAHYPERBOLIC METRICS

Quite generally, a Riemannian metric given by the fundamental form

$$ds^2 = \rho^2(dx^2 + dy^2), \tag{1-20}$$

or $ds = \rho|dz|$, $\rho > 0$, is conformal with the euclidean metric. The quantity

$$K(\rho) = -\rho^{-2} \Delta \log \rho$$

is known as the *curvature* (or gaussian curvature) of the metric (1-20). The reader will verify that the metrics (1-5) and (1-6) have constant curvature -1 [the factor 2 in (1-5) was chosen with this in mind].

In this text, which deals primarily with complex variables, the geometric definition of curvature is unimportant, and we use the name only as a convenience. It is essential, however, that $K(\rho)$ is invariant under conformal mappings.

Consider a conformal mapping $w = f(z)$ and define $\bar{\rho}(w)$ so that $\rho|dz| = \bar{\rho}|dw|$ or, more explicitly, $\rho(z) = \bar{\rho}[f(z)]|f'(z)|$. Because $\log |f'(z)|$ is harmonic, it follows that $\Delta \log \rho(z) = \Delta \log \bar{\rho}(w)$, both laplacians being with respect to z. Change of variable in the laplacian follows the rule $\Delta_z \log \tilde{\rho} = |f'(z)|^2 \Delta_w \log \tilde{\rho}$, and we find that $K(\rho) = K(\tilde{\rho})$.

From now on the hyperbolic metric in Δ will be denoted by $\lambda|dz|$; that is to say, we set

$$\lambda(z) = \frac{2}{1 - |z|^2}.$$

We wish to compare $\lambda|dz|$ with other metrics $\rho|dz|$.

Lemma 1-1   If ρ satisfies $K(\rho) \leq 1$ everywhere in Δ, then $\lambda(z) \geq \rho(z)$ for all $z \in \Delta$.

PROOF   We assume first that ρ has a continuous and strictly positive extension to the closed disk. From $\Delta \log \lambda = \lambda^2$, $\Delta \log \rho \geq \rho^2$ we have $\Delta(\log \lambda - \log \rho) \leq \lambda^2 - \rho^2$. The function $\log \lambda - \log \rho$ tends to $+\infty$ when $|z| \to 1$. It therefore has a minimum in the unit disk. At the point of minimum $\Delta(\log \lambda - \log \rho) \geq 0$ and hence $\lambda^2 \geq \rho^2$, proving that $\lambda \geq \rho$ everywhere.

To prove the lemma in the general case we replace $\rho(z)$ by $r\rho(rz)$, $0 < r < 1$. This metric has the same curvature, and the smoothness condition is fulfilled. Hence $\lambda(z) \geq r\rho(rz)$, and $\lambda(z) \geq \rho(z)$ follows by continuity.

The definition of curvature requires $\Delta \log \rho$ to exist, so we have to assume that ρ is strictly positive and of class C^2. These restrictions are inessential and cause difficulties in the applications. They can be removed in a way that is reminiscent of the definition of subharmonic functions.

Definition 1-1   A metric $\rho|dz|$, $\rho \geq 0$ is said to be ultrahyperbolic in a region Ω if it has the following properties:

(*i*) ρ is upper semicontinuous.
(*ii*)   At every $z_0 \in \Omega$ with $\rho(z_0) > 0$ there exists a "supporting metric" ρ_0, defined and of class C^2 in a neighborhood V of z_0, such that $\Delta \log \rho_0 \geq \rho_0^2$ and $\rho \geq \rho_0$ in V, while $\rho(z_0) = \rho_0(z_0)$.

Because $\log \lambda - \log \rho$ is lower semicontinuous, the existence of a minimum is still assured. The minimum will also be a local minimum of $\log \lambda - \log \lambda_0$, and the rest of the reasoning applies as before. The inequality $\lambda(z) \geq \rho(z)$ holds as soon as ρ is ultrahyperbolic.

We are now ready to prove a stronger version of Schwarz's lemma.

Theorem 1-7   Let f be an analytic mapping of Δ into a region Ω on which there is given an ultrahyperbolic metric ρ. Then $\rho[f(z)]|f'(z)| \leq 2(1 - |z|^2)^{-1}$.

The proof consists in the trivial observation that $\rho[f(z)]|f'(z)|$ is ultrahyperbolic on Δ. Observe that the zeros of $f'(z)$ are singularities of this metric.

REMARK   The notion of an ultrahyperbolic metric makes sense, and the theorem remains valid if Ω is replaced by a Riemann surface. In this book only the last two chapters deal systematically with Riemann surfaces, but we shall not hesitate to make occasional references to Riemann surfaces when the need arises. Thus in our next section we shall meet an application of Theorem 1-7 in which Ω is in fact a Riemann surface, but the adaptation will be quite obvious.

1-6   BLOCH'S THEOREM

Let $w = f(z)$ be analytic in Δ and norm lized by $|f'(0)| = 1$. We may regard f as a one-to-one mapping of Δ onto a Riemann surface W_f spread over the w plane. It is intuitively clear what is meant by an unramified disk contained in W_f. As a formal definition we declare that an unramified disk is an open disk Δ' together with an open set $D \subset \Delta$ such that f restricted to D defines a one-to-one mapping of D onto Δ'. Let B_f denote the least upper bound of the radii of all such disks Δ'. Bloch made the important observation that B_f cannot be arbitrarily small. In other words, the greatest lower bound of B_f for all normalized f is a positive number B, now known as Bloch's constant. Its value is not known, but we shall prove Theorem 1-8:

Theorem 1-8 $B \geq \sqrt{3}/4.$

PROOF   Somewhat informally we regard $w = f(z)$ both as a point on W_f and as a complex number. Let $R(w)$ be the radius of the largest unramified disk of center w contained in W_f [at a branch-point $R(w) = 0$]. We introduce a metric $\bar{\rho}|dw|$ on W_f defined by

$$\bar{\rho}(w) = \frac{A}{R(w)^{\frac{1}{2}}[A^2 - R(w)]}$$

where A is a constant $> B_f^{\frac{1}{2}}$. This induces a metric $\rho(z) = \bar{\rho}[f(z)]|f'(z)|$ in Δ. We wish to show that $\rho(z)$ is ultrahyperbolic for a suitable choice of A.

Suppose that the value $w_0 = f(z_0)$ is assumed with multiplicity $n > 1$. For w close to w_0 (or rather z close to z_0), $R(w) = |w - w_0|$, which is of the order $|z - z_0|^n$. Since $|f'(z)|$ is of order $|z - z_0|^{n-1}$, it follows that $\rho(z)$ is of order $|z - z_0|^{n/2-1}$. If $n > 2$, it follows that ρ is continuous and

$\rho(z_0) = 0$. We recall that there is no need to look for a supporting metric at points where ρ is zero.

In case $n = 2$ we have

$$\rho(z) = \frac{A|f'(z)|}{|f(z) - f(z_0)|^{\frac{1}{2}}[A^2 - |f(z) - f(z_0)|]}$$

near z_0. This metric is actually regular at z_0, and it satisfies $\Delta \log \rho = \rho^2$ as seen either by straightforward computation or from the fact that $\rho|dz| = 2|dt|/(1 - |t|^2)$ with $t = A^{-1}[f(z) - f(z_0)]^{\frac{1}{2}}$.

It remains to find a supporting metric at a point $w_0 = f(z_0)$ with $f'(z_0) \neq 0$. Denote the disk $\{w; |w - w_0| < R(w_0)\}$ by $\Delta'(w_0)$ and by $D(z_0)$ the component of its inverse image that contains z_0. The boundary of $D(z_0)$ must contain either a point $a \in \Delta$ with $f'(a) = 0$, or a point a on the unit circle, for otherwise $\Delta'(w_0)$ would not be maximal. In the first case the boundary of $\Delta'(w_0)$ passes through the branch-point $b = f(a)$. In the second case $f(a)$ is not defined, but we make the harmless assumption that f can be extended continuously to the closed unit disk. The point $b = f(a)$ is then on the boundary of $\Delta'(w_0)$ and may also be regarded as a boundary point of the Riemann surface W_f.

Choose $z_1 \in D(z_0)$, $w_1 = f(z_1) \in \Delta'(w_0)$. It is geometrically clear that $R(w_1) \leq |w_1 - b|$. For a more formal reasoning we consider $\Delta'(w_1)$ and $D(z_1)$. Let c be the line segment from w to b. If b were in $\Delta'(w_1)$, all of c except the last point would be in $\Delta'(w_0) \cap \Delta'(w_1)$. But the inverse functions f^{-1} with values in $D(z_0)$ and $D(z_1)$ agree on this set, and it would follow by continuity that $a \in D(z_1)$. This is manifestly impossible. We conclude that b is not in $\Delta'(w_1)$, and hence that $R(w_1) \leq |w_1 - b|$.

Now we compare $\rho(z)$ with

$$\rho_0(z) = \frac{A|f'(z)|}{|f(z) - b|^{\frac{1}{2}}[A^2 - |f(z) - b|]}$$

when z is close to z_0. This metric has constant curvature -1 and

$$\rho_0(z_0) = \rho(z_0).$$

Moreover, the inequality $\rho(z) \geq \rho_0(z)$ holds near z_0 if the function $t^{\frac{1}{2}}(A^2 - t)$ remains increasing for $0 \leq t \leq R(w_0)$. The derivative changes sign at $t = A^2/3$. We conclude that $\rho(z)$ is ultrahyperbolic if $A^2 > 3B_f$.

All that remains is to apply Lemma 1-1 with $z = 0$. We obtain $A \leq 2R[f(0)]^{\frac{1}{2}}\{A^2 - R[f(0)]\} \leq 2B_f^{\frac{1}{2}}(A^2 - B_f)$. The inequality $B_f \geq \sqrt{3}/4 > 0.433$ follows on letting A tend to $(3B_f)^{\frac{1}{2}}$.

It is conjectured that the correct value of B is approximately 0.472. This value is assumed for a function that maps Δ on a Riemann surface with branch points of order 2 over all vertices in a net of equilateral triangles.

1-7 THE POINCARÉ METRIC OF A REGION

The hyperbolic metric of a disk $|z| < R$ is given by

$$\lambda_R(z) = \frac{2R}{R^2 - |z|^2}.$$ (1-21)

If ρ is ultrahyperbolic in $|z| < R$, we must have $\rho \leq \lambda_R$. In particular, if ρ were ultrahyperbolic in the whole plane we would have $\rho = 0$. Hence there is no ultrahyperbolic metric in the whole plane.

The same is true of the punctured plane $\{z; z \neq 0\}$. Indeed, if $\rho(z)$ were ultrahyperbolic in the punctured plane, then $\rho(e^z)|e^z|$ would be ultrahyperbolic in the full plane. These are the only cases in which an ultrahyperbolic metric fails to exist.

Theorem 1-9 In a plane region Ω whose complement has at least two points, there exists a unique maximal ultrahyperbolic metric, and this metric has constant curvature -1.

The maximal metric is called the *Poincaré metric* of Ω, and we denote it by λ_Ω. It is maximal in the sense that every ultrahyperbolic metric ρ satisfies $\rho \leq \lambda_\Omega$ throughout Ω. The uniqueness is trivial.

The existence proof is nonelementary and will be postponed to Chap. 10. The reader will note, however, that the applications we are going to make do not really depend on the existence of the Poincaré metric. At present its main purpose is to allow a convenient terminology.

Theorem 1-10 If $\Omega \subset \Omega'$, then $\lambda_{\Omega'} \leq \lambda_\Omega$.

This is obvious, for the restriction of $\lambda_{\Omega'}$ to Ω is ultrahyperbolic in Ω.

Theorem 1-11 Let $\delta(z)$ denote the distance from $z \in \Omega$ to the boundary of Ω. Then $\lambda_\Omega(z) \leq 2/\delta(z)$.

Ω contains the disk with center z and radius $\delta(z)$. The estimate follows from Theorem 1-10 together with (1-21). It is the best possible, for equality holds when Ω is a disk and z its center.

It is a much harder problem to find lower bounds.

1-8 AN ELEMENTARY LOWER BOUND

Let $\Omega_{a,b}$ be the complement of the two-point set $\{a,b\}$ and denote its Poincaré metric by $\lambda_{a,b}$. If a and b are in the complement of Ω, then

$\Omega \subset \Omega_{a,b}$ and $\lambda_\Omega \geq \lambda_{a,b}$. A lower bound for $\lambda_{a,b}$ is therefore a lower bound for λ_Ω. Because

$$\lambda_{a,b}(z) \ = \ |b \ - \ a|^{-1}\lambda_{0,1}\frac{(z \ - \ a)}{(b \ - \ a)}$$

it is sufficient to consider $\lambda_{0,1}$. There are known analytic expressions for $\lambda_{0,1}$, but they are not of great use. What we require is a good elementary lower bound.

The region $\Omega_{0,1}$ is mapped on itself by $1 - z$ and by $1/z$. Therefore $\lambda_{0,1}(z) \ = \ \lambda_{0,1}(1 \ - \ z) \ = \ |z|^{-2}\lambda_{0,1}(1/z)$. It follows that we need consider only $\lambda_{0,1}$ in one of the regions $\Omega_1, \Omega_2, \Omega_3$ marked in Fig. 1-1.

We begin by determining a better upper bound than the one given by Theorem 1-11. $\Omega_{0,1}$ contains the punctured disk $0 < |z| < 1$. The Poincaré metric of the punctured disk is found by mapping its universal covering, an infinitely many-sheeted disk, on the half plane Re $w < 0$ by means of $w = \log z$. The metric is $|dw|/|\mathrm{Re}\ w| = |dz|/|z| \log (1/|z|)$, and we obtain

$$\lambda_{0,1}(z) \ \leq \ \left(|z| \log \frac{1}{|z|}\right)^{-1} \tag{1-22}$$

for $|z| < 1$. This estimate shows what order of magnitude to expect.

Let $\zeta(z)$ be the function that maps the complement of $[1, +\infty]$ conformally on the unit disk, origins corresponding to each other and symmetry with respect to the real axis being preserved.

Theorem 1-12   For $|z| \leq 1$, $|z| \leq |z - 1|$, i.e., for $z \in \Omega_1$,

$$\lambda_{0,1}(z) \ \geq \ \left|\frac{\zeta'(z)}{\zeta(z)}\right| [4 \ - \ \log |\zeta(z)|]^{-1}. \tag{1-23}$$

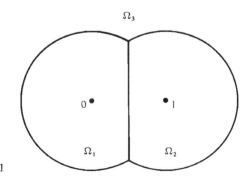

FIGURE 1-1

For $z \to 0$, (1-22) and (1-23) imply

$$\log \lambda_{0,1}(z) = -\log |z| - \log \log \frac{1}{|z|} + O(1). \qquad (1\text{-}24)$$

PROOF It is immediate that the metric defined by

$$\rho(z) = \left| \frac{\zeta'(z)}{\zeta(z)} \right| [4 - \log |\zeta(z)|]^{-1} \qquad (1\text{-}25)$$

has curvature -1, for it is obtained from the Poincaré metric of the punctured disk $0 < |\zeta| < e^4$. We use (1-25) only in Ω_1 and extend ρ to Ω_2 and Ω_3 by means of the symmetry relations $\rho(1 - z) = \rho(z)$ and $\rho(1/z) = |z|^2 \rho(z)$. The extended metric is obviously continuous. We need to verify that ρ has a supporting metric on the lines that separate $\Omega_1, \Omega_2, \Omega_3$. Because of the symmetry it is sufficient to consider the line segment between Ω_1 and Ω_2. It is readily seen that the original ρ, as given by (1-25) in Ω_1 and part of Ω_2, constitutes a supporting metric provided $\partial \rho / \partial x < 0$ on the separating line segment.

The mapping function is given explicitly by

$$\zeta(z) = \frac{\sqrt{1 - z} - 1}{\sqrt{1 - z} + 1}$$

with Re $\sqrt{1 - z} > 0$. In

$$\frac{\partial \log \rho}{\partial x} = \text{Re} \left(\frac{d}{dz} \log \frac{\zeta'}{\zeta} \right) + \text{Re} \frac{\zeta'}{\zeta} (4 - \log |\zeta|)^{-1}$$

we substitute

$$\frac{\zeta'}{\zeta} = \frac{1}{z \sqrt{1 - z}}$$

$$\frac{d}{dz} \log \frac{\zeta'}{\zeta} = \frac{3z - 2}{2z(1 - z)}.$$

On taking into account that $1 - z = \bar{z}$ on the line segment, we find

$$\frac{\partial \log \rho}{\partial x} = -\frac{1}{4|z|^2} + \frac{\text{Re} \sqrt{z}}{|z|^2} (4 - \log |\zeta|)^{-1},$$

and this is negative because $|\zeta| < 1$ and Re $\sqrt{z} < 1$.

We conclude that (1-23) holds. The passage to (1-24) is a trivial verification.

1-9   THE PICARD THEOREMS

We use Theorems 1-7 and 1-12 to prove a classic theorem known as the Picard-Schottky theorem. The emphasis is on the elementary nature of the proof and the explicit estimates obtained.

Theorem 1-13   Suppose that $f(z)$ is analytic and different from 0 and 1 for $|z| < 1$. Then

$$\log |f(z)| \leq [7 + \overset{+}{\log} |f(0)|] \frac{1 + |z|}{1 - |z|}. \tag{1-26}$$

REMARK   As usual, $\overset{+}{\log} |f(0)|$ is the greater of $\log |f(0)|$ and 0. The constant in the bound is not the best possible, but the order of magnitude of the right-hand side is right.

PROOF   Because $1/f$ satisfies the same conditions as f it is irrelevant whether we derive an upper or a lower bound for $\log |f|$. The way we have formulated Theorem 1-12, it is slightly more convenient to look for a lower bound.

By assumption f maps Δ into $\Omega_{0,1}$. By Theorem 1-7 we therefore have

$$\lambda_{0,1}[f(z)]|f'(z)| \leq \frac{2}{1 - |z|^2}.$$

We obtain by integration

$$\int_{f(0)}^{f(z)} \lambda_{0,1}(w)|dw| \leq \log \frac{1 + |z|}{1 - |z|}, \tag{1-27}$$

where the integral is taken along the image of the line segment from 0 to z. We use the notation Ω_1 of the previous section and assume first that the whole path of integration lies in Ω_1. The estimate (1-23) can be applied and gives

$$\int_{f(0)}^{f(z)} (4 - \log |\zeta(w)|)^{-1}|d \log \zeta(w)| \leq \log \frac{1 + |z|}{1 - |z|}. \tag{1-28}$$

On noting that $|d \log \zeta| \geq -d \log |\zeta|$ we find

$$\frac{4 - \log |[f(z)]|}{4 - \log |[f(0)]|} \leq \frac{1 + |z|}{1 - |z|}. \tag{1-29}$$

From the explicit expression

$$|\zeta(w)| = \frac{|w|}{|1 + \sqrt{1 - w}|^2}$$

we derive $(1 + \sqrt{2})^{-2}|w| \leq |\zeta(w)| \leq |w|$, the lower bound being quite crude. With these estimates, and since $\log(1 + \sqrt{2}) < 1$, we obtain from (1-29)

$$- \log|f(z)| < [6 - \log|f(0)|] \frac{1 + |z|}{1 - |z|}. \tag{1-30}$$

Now let us drop the assumption that the path in (1-27) stays in Ω_1. If $f(z) \in \Omega_1$, (1-28) is still true if we start the integral from w_0, the last point on the boundary of Ω_1. Since $|w_0| \geq \frac{1}{2}$, the inequality (1-30) is replaced by

$$-\log|f(z)| < (6 + \log 2) \frac{1 + |z|}{1 - |z|} \tag{1-31}$$

which is also trivially true in case $f(z)$ is not in Ω_1. The inequalities (1-30) and (1-31) can be combined to give

$$-\log|f(z)| < \left[6 + \log 2 + \overset{+}{\log} \frac{1}{|f(0)|} \right] \frac{1 + |z|}{1 - |z|},$$

and (1-26) is a weaker version with f replaced by $1/f$. The theorem is proved.

Corollary   The little Picard theorem   If f is meromorphic in the whole plane and omits three values, then f is constant.

PROOF   If f omits a,b,c then $F = [(c - b)/(c - a)][(f - a)/(f - b)]$ is holomorphic and omits 0,1. Apply Theorem 1-13 to $F(Rz)$ with $R > 0$. It follows that $|F(Re^{i\theta}/2)|$ lies under a finite bound, independent of R and θ. Hence $|F(z)|$ is bounded, and F must be a constant by Liouville's theorem.

Theorem 1-14   The big Picard theorem   If f is meromorphic and omits three values in a punctured disk $0 < |z| < \delta$, then it has a meromorphic extension to the full disk.

PROOF   We may assume that $\delta = 1$ and that f omits $0, 1, \infty$. Comparison of $\lambda_{0,1}$ with the Poincaré metric of the punctured disk yields

$$\lambda_{0,1}[f(z)]|f'(z)| \leq \left(|z| \log \frac{1}{|z|} \right)^{-1}.$$

We integrate along a radius from $z_0 = r_0 e^{i\theta}$ to $z = re^{i\theta}$, $r < r_0 < 1$. If $f(z) \in \Omega_1$, we obtain as in the preceding proof

$$\log \{4 - \log|f(z)|\} \leq \log \log \frac{1}{|z|} + A,$$

where A is an irrelevant constant. This implies

$$-\log |f(z)| \leq C \log \frac{1}{|z|}$$

with some other constant, showing that $1/|f|$ is bounded by a power of $1/|z|$. Hence the isolated singularity at the origin is not essential.

NOTES   The Schwarz lemma and its classic proof are due to Carathéodory [10]; Schwarz proved it only for one-to-one mappings [58, p. 109]. Although Poincaré had used noneuclidean geometry for function theoretic purposes, Pick [50, 51] seems to be the first to have fully realized the invariant character of Schwarz's lemma. Theorem 1-2 has been included mainly for historical reasons.

Theorem 1-5 was first proved by Carathéodory [11] but independently and almost simultaneously by Landau and Valiron [35]. All three were unaware that the theorem is an easy consequence of Herglotz's integral representation of positive harmonic functions. We have given preference to Carathéodory's proof because of its geometric character.

Ultrahyperbolic metrics (without the name) were introduced by Ahlfors [1]. They have recently found many new applications in the theory of several complex variables.

There are many proofs of Bloch's theorem, that of Landau [34] probably being the simplest. The original theorem is in Bloch [8]. Heins has improved on the author's bound by showing that $B > \sqrt{3}/4$ (Heins [28]). See also Pommerenke [52].

Stronger forms of (1-26) can be found in Jenkins [32], but his proof uses the modular function. Our proof of the Picard theorems is elementary not only because it avoids the modular function, but also because it does not use the monodromy theorem.

EXERCISES

1   Derive formulas for the noneuclidean center and radius of a circle contained in the unit disk or the half plane.

2   Show that two circular arcs in the unit disk with common end points on the unit circle are noneuclidean parallels in the sense that the points on one arc are at constant distance from the other.

3   Let $z = z(t)$ be an arc of class C^3. Show that the rate of change of its curvature can be expressed through

$$|z'(t)|^{-1} \operatorname{Im} \left[\frac{z'''(t)}{z'(t)} - \frac{3}{2} \left(\frac{z''(t)}{z'(t)} \right)^2 \right].$$

4 Formulate and prove the analog of Theorem 1-5 for functions with positive real part on the right half plane.

5 Verify that the spherical metric

$$ds = \frac{2|dz|}{1 + |z|^2}$$

has constant curvature 1.

6 If f is analytic in the unit disk Δ and normalized by $|f'(0)| = 1$, let L_f be the least upper bound of the radii of all disks covered by the image $f(\Delta)$. Imitating the proof of Bloch's theorem, show that the greatest lower bound of L_f is a constant $L \geq \frac{1}{2}$.

2

CAPACITY

2-1 THE TRANSFINITE DIAMETER

Let E be a closed bounded set in the complex plane. We define its diameter of order n as

$$d_n = \max \prod_{i<j} |z_i - z_j|^{2/n(n-1)}$$

for points $z_i \in E$, $i = 1, \ldots, n$. If z_k is omitted,

$$\prod_{i,j \neq k} |z_i - z_j| \leq d_{n-1}^{(n-1)(n-2)/2}.$$

When these inequalities are multiplied, each factor $|z_i - z_j|$ occurs $n - 2$ times, and we obtain

$$d_n^{n(n-1)(n-2)/2} \leq d_{n-1}^{n(n-1)(n-2)/2},$$

and hence $d_n \leq d_{n-1}$. We set $d_\infty = \lim d_n$ and call it the *transfinite diameter* of E.

Among all monic polynomials $P_n(z) = z^n + a_1 z^{n-1} + \cdots + a_n$ of degree n, there is one whose maximum modulus on E is a minimum. It is called a Chebyshev polynomial, and we denote its maximum modulus on E by $\rho_n{}^n$.

Theorem 2-1 $\quad \lim_{n \to \infty} \rho_n = d_\infty$.

PROOF Let $z_1, \ldots, z_n$ be the end points of d_n and consider the Vandermonde determinant

$$
V(z, z_1, \ldots, z_n) = \begin{vmatrix} 1 & z & \cdots & z^n \\ 1 & z_1 & \cdots & z_1{}^n \\ \cdots & \cdots & \cdots & \cdots \\ 1 & z_n & \cdots & z_n{}^n \end{vmatrix}.
$$

It is a polynomial whose highest coefficient has absolute value $d_n{}^{n(n-1)/2}$. The maximum of $|V|$ on E is $\leq d_{n+1}{}^{n(n+1)/2} \leq d_n{}^{n(n+1)/2}$. It follows that $\rho_n{}^n \leq d_n{}^n$, $\rho_n \leq d_n$.

Next we observe that

$$
\begin{vmatrix} 1 & \cdots & z_1{}^{n-1} \\ \cdots & \cdots & \cdots \\ 1 & \cdots & z_n{}^{n-1} \end{vmatrix} = \begin{vmatrix} 1 & P_1(z_1) & \cdots & P_{n-1}(z_1) \\ \cdots & \cdots & \cdots & \cdots \\ 1 & P_1(z_n) & \cdots & P_{n-1}(z_n) \end{vmatrix},
$$

where the P_k may be chosen as Chebyshev polynomials. On using the Hadamard inequality for determinants, we obtain $d_n{}^{n(n-1)/2} \leq n^{n/2} \rho_1 \rho_2{}^2 \cdots \rho_{n-1}{}^{n-1}$, and hence $\liminf (\rho_1 \rho_2{}^2 \cdots \rho_{n-1}{}^{n-1})^{2/n(n-1)} \geq d_\infty$. This is a weighted geometric mean, and we conclude that if $\lim \rho_n$ exists it must be equal to d_∞.

To prove the existence of the limit we use the inequality $\rho_{mk+h}^{mk+h} \leq \max |P_m{}^k P_h| \leq \rho_m{}^{mk} \rho_h{}^h$ which we write as

$$
\log \rho_{mk+h} \leq \frac{mk}{mk+h} \log \rho_m + \frac{h}{mk+h} \log \rho_h.
$$

Keep m fixed and let k run through the positive integers while $h = 0$, $\ldots$, $m-1$. We conclude that $\limsup \rho_n \leq \rho_m$, which obviously implies the existence of $\lim \rho_n$.

2-2 POTENTIALS

Consider a positive mass distribution μ on the compact set E, i.e., a measure that vanishes on the complement of E. We define

$$
p_N(z) = \int \min\left(N, \log \frac{1}{|z - \zeta|}\right) d\mu(\zeta)
$$

and $p(z) = \lim_{N \to \infty} p_N(z)$. This is the *logarithmic potential* of μ. Clearly, p is lower semicontinuous, $p(z_0) \leq \lim \inf_{z \to z_0} p(z)$, and harmonic outside of E. We set $V_\mu = \sup_z p(z)$. It may be infinite.

If ν is another mass distribution, we can form

$$I(\mu,\nu) = \lim_{N \to \infty} \int p_N(z) \, d\nu(z).$$

We leave it to the reader to prove that $I(\mu,\nu) = I(\nu,\mu)$. For $\mu = \nu$ we write $I(\mu)$ instead of $I(\mu,\mu)$. It is the *energy integral* of μ.

Theorem 2-2   Among all distributions with total mass $\mu(E) = 1$, there is one that minimizes V_μ. The same distribution minimizes $I(\mu)$, and the two minima are equal.

Definition 2-1   If min $V_\mu = V$, we call e^{-V} the *capacity* of E.

REMARK   It may happen that $V = \infty$, namely, if no μ gives rise to a finite V_μ. Then E is a set of zero capacity.

PROOF   The proof of Theorem 2-2 is in several steps. We assume first that the complement of E is connected and bounded by a finite number of piecewise analytic Jordan curves. We denote the complement by Ω and its boundary by $\partial\Omega$. The orientation of $\partial\Omega$ is chosen so that Ω lies to the left.

It is known that Ω has a Green's function with a pole at ∞ (see Ahlfors, L. V.: "Complex Analysis," 2d ed., McGraw-Hill Book Company, New York, 1966, henceforth referred to as C.A.). The Green's function is harmonic in Ω, it vanishes on $\partial\Omega$, and its asymptotic behavior at ∞ is of the form

$$g(z) = \log |z| + \gamma + \epsilon(z),$$

where γ is a constant and $\epsilon(z) \to 0$ for $z \to \infty$. The constant γ is known as the *Robin constant*.

For any $\zeta \in \Omega$, Green's formula yields

$$g(\zeta) - \gamma = \frac{1}{2\pi} \int_{\partial\Omega} \log \frac{1}{|z - \zeta|} \frac{\partial g}{\partial n} |dz|, \qquad (2\text{-}1)$$

where the normal derivative is in the direction of the outer normal (we adopt this convention throughout this book). It is clear that $\partial g/\partial n < 0$, and we can define a positive mass distribution by setting

$$\mu(e) = -\frac{1}{2\pi} \int_{e \cap \partial\Omega} \frac{\partial g}{\partial n} |dz|$$

for any Borel set e. Green's formula shows that the total mass is 1.

Formula (2-1) shows that the potential of μ satisfies $p(\zeta) = \gamma - g(\zeta)$ for $\zeta \in \Omega$. Green's formula can also be applied when ζ is an exterior point of Ω, and even if $\zeta \in \partial\Omega$. We find that $p(\zeta) = \gamma$ on E. Hence $V_\mu = \gamma$, and we have proved that $V \leq \gamma$.

Let μ_0 be another mass distribution with total mass 1 and let p_0 be its potential. Then $p_0(z) - p(z) \to 0$ for $z \to \infty$, and it follows by the maximum principle that $V_{\mu_0} \geq V_\mu = \gamma$. Thus V_μ is minimal, and $V = \gamma$. It follows further that

$$I(\mu,\mu_0) = \int p \, d\mu_0 = \gamma = I(\mu). \tag{2-2}$$

To continue the proof we need a lemma.

Lemma 2-1   Let μ_1 and μ_2 be positive mass distributions on E with $\mu_1(E) = \mu_2(E)$ and $I(\mu_1) < \infty$, $I(\mu_2) < \infty$. Then $I(\mu_1) + I(\mu_2) - 2I(\mu_1,\mu_2) \geq 0$.

PROOF   It is elementary to show that

$$\frac{1}{2\pi} \iint\limits_{|z|<R} \frac{dx \, dy}{|z - z_1| \, |z - z_2|} = \log R - \log |z_1 - z_2| + C + \epsilon(z_1,z_2,R),$$

$$\tag{2-3}$$

where C is a constant and $\epsilon(z_1,z_2,R) \to 0$ for $R \to \infty$, uniformly when z_1,z_2 are on a compact set. We may assume that $\mu_1(E) = \mu_2(E) = 1$. Integration of (2-3) with respect to $\mu_i(z_1)$ and $\mu_j(z_2)$, $i,j = 1,2$, yields

$$\frac{1}{2\pi} \iint\limits_{|z|<R} \left[\int \frac{d\mu_i(\zeta)}{|\zeta - z|} \int \frac{d\mu_j(\zeta)}{|\zeta - z|} \right] dx \, dy = \log R + I(\mu_i,\mu_j) + C + \epsilon(R)$$

with $\epsilon(R) \to 0$. It follows that

$$I(\mu_1) + I(\mu_2) - 2I(\mu_1,\mu_2) = \lim_{R\to\infty} \iint\limits_{|z|<R} \left[\int \frac{d\mu_1(\zeta) - d\mu_2(\zeta)}{|\zeta - z|} \right]^2 dx \, dy \geq 0.$$

Another way of expressing the result is to state that $I(\mu_1 - \mu_2) \geq 0$.

We apply the lemma to μ and μ_0. It follows from (2-2) together with the lemma that $I(\mu_0) \geq I(\mu)$. We have proved that $I(\mu)$ is minimal. The distribution μ is known as the equilibrium distribution.

It remains to pass to the case of an arbitrary compact set E. The unbounded component of the complement of E is denoted by Ω. It can be represented as the union of a sequence of increasing regions Ω_n each of which satisfies our earlier conditions. The complement of Ω_n will be denoted by E_n, the equilibrium distribution on E_n by μ_n, the potential of μ_n by

p_n, the Green's function by g_n, and the Robin constant by γ_n. By the maximum principle g_n and γ_n increase with n. We set $g(z) = \lim_{n\to\infty} g_n(z)$ and $\gamma = \lim \gamma_n$. By Harnack's principle $g(z)$ is either harmonic or identically $+\infty$. If it is finite, g is called the Green's function of Ω; it is easily seen to be independent of the sequence $\{\Omega_n\}$ by which it is defined.

It is well known that one can select a subsequence of the μ_n that converges to a limit distribution μ with the same total mass. Evidently, μ is a distribution on E, and in fact on the boundary of Ω. For convenience we adjust the notation so that $\{\mu_n\}$ is the subsequence.

If z is not on the boundary of Ω, it is immediate that the potential p of μ satisfies

$$p(z) = \lim p_n(z) \leq \gamma.$$

Because of the lower semicontinuity this inequality remains true on the boundary, and we conclude that $V_\mu \leq \gamma$. On the other hand, if μ_0 is any distribution of unit mass on E, it is also a distribution on E_n so that $V_{\mu_0} \geq \gamma_n$, and hence $V_{\mu_0} \geq \gamma$. We have shown that V_μ is a minimum and equal to γ.

It can no longer be asserted that $p(z)$ is constantly equal to γ on E. However, it is trivial that $I(\mu) \leq \gamma$, and for any distribution μ_0 of unit mass on E we have $I(\mu_0) \geq I(\mu_n) = \gamma_n$. Hence $I(\mu_0) \geq \gamma$ and, in particular, $I(\mu) \geq \gamma$, so that in fact $I(\mu) = \gamma$. We have proved that $I(\mu)$ is indeed a minimum and equal to the minimum of V_u.

2-3 CAPACITY AND THE TRANSFINITE DIAMETER

We proved in the preceding section that cap $E = e^{-\gamma}$, where γ is the Robin constant of Ω, the unbounded component of the complement of E. In particular, the capacity does not change if E is replaced by the full complement of Ω.

It is clear that γ, and therefore the capacity of E, has a certain degree of invariance with respect to conformal mappings of Ω. In fact, suppose that $f(z)$ defines a conformal mapping of Ω on a region Ω_1, and that the Laurent development of $f(z)$ at ∞ has the form $f(z) = z + \cdots$ so that $f(\infty) = \infty$ and $f(z)/z \to 1$. If g_1 is the Green's function of Ω_1, then $g_1 \circ f$ is the Green's function of Ω. The Robin constants γ and γ_1 are equal. Hence the capacity of E_1, the complement of Ω_1, is equal to that of E. In other words, the capacity is invariant under normalized conformal mappings.

Note that there is no mapping of E on E_1; the comparison comes about by passing to the complements. If we drop the normalization, we have cap $E_1 = |a|$ cap E, where $f(z) = az + \cdots$. A quantity with this behavior may be called a *relative conformal invariant*.

The capacity of a disk of radius R is R. The capacity of a line segment of length L is $L/4$.

We shall now study capacity in its relation to the transfinite diameter.

Theorem 2-3   The capacity of a closed bounded set is equal to its transfinite diameter.

PROOF   With the same notations as before, let μ be the equilibrium distribution and $P_n(z) = (z - \zeta_1) \cdots (z - \zeta_n)$ the Chebyshev polynomial of degree n. It is immediate by Green's formula that

$$\int \log |P_n| \, d\mu = -p(\zeta_1) - \cdots - p(\zeta_n) \geq -n\gamma.$$

Hence $\rho_n{}^n = \max_E |P_n| \geq e^{-n\gamma}$, and it follows that $d_\infty \geq e^{-\gamma} = $ cap E.

For the opposite inequality we observe that $d_\infty(E) \leq d_\infty(E_n)$. Hence if we prove that $d_\infty(E_n) \leq e^{-\gamma_n}$, it will follow that $d_\infty(E) \leq e^{-\gamma}$. In other words, we are free to assume that Ω has analytic boundary curves.

We divide the boundary $\partial\Omega$ into n parts c_i such that each c_i carries exactly the mass $1/n$ of the equilibrium distribution. For large n most of the parts can be chosen as arcs, but if there are N contours we must allow for $N - 1$ parts which are not connected. These parts will be called exceptional.

We choose points $\zeta_i \in c_i$ and consider the polynomial

$$P_n(z) = (z - \zeta_1) \cdots (z - \zeta_n).$$

Recall that the potential of the equilibrium distribution equals γ on E. Since the mass on each c_i is $1/n$, we obtain

$$\frac{1}{n} \log |P_n(z)| + \gamma = \sum_i \int_{c_i} \log \frac{|z - \zeta_i|}{|z - \zeta|} \, d\mu(\zeta)$$

for all $z \in E$. We can choose n so large that the diameter of each non-exceptional c_i is less than a fixed $\delta > 0$. For $z \in c_i$ we then have

$$\log \left| \frac{z - \zeta_i}{z - \zeta} \right| = \log \left| 1 + \frac{\zeta - \zeta_i}{z - \zeta} \right| \leq \log \left(1 + \frac{\delta}{|z - \zeta|} \right).$$

When z lies on an exceptional part we can only say that

$$\log \left| \frac{z - \zeta_i}{z - \zeta} \right| \leq \log \frac{D}{|z - \zeta|},$$

where D is the diameter of E. With these estimates we obtain

$$\frac{1}{n} \log |P_n(z)| + \gamma \le \int \log \left(1 + \frac{\delta}{|z - \zeta|} \right) d\mu + \int_{c'} \log \frac{D}{|z - \zeta|} \, d\mu, \quad (2\text{-}4)$$

where c' is the union of the exceptional parts.

Let d be the shortest distance between contours. We obtain trivial estimates for the contours that do not contain z, the total contribution to the right-hand side of (2-4) being at most

$$\log \left(1 + \frac{\delta}{d} \right) + \frac{N-1}{n} \log \frac{D}{d}. \quad (2\text{-}5)$$

Denote by s the arc length from z to ζ along the contour that contains z. Because of the regularity of the contour there exists a constant $k > 0$ such that $|z - \zeta| \ge ks$. Moreover, the normal derivative of g is bounded so that $d\mu \le K ds$ with finite K. It is seen that the remaining part of the first integral in (2-4) is bounded by

$$2K \int_0^{L/2} \log \left(1 + \frac{\delta}{ks} \right) ds, \quad (2\text{-}6)$$

where L is the length of the contour. The remaining part of the second integral is at most equal to

$$K \int_0^{\delta/k} \log \frac{D}{ks} \, ds. \quad (2\text{-}7)$$

Recall that $\rho_n{}^n \le \max |P_n(z)|$ on E, that is to say, on $\partial\Omega$. Our estimates show that $\log \rho_n + \gamma$ is bounded by the sum of the expressions (2-5) to (2-7). All three tend to zero when $n \to \infty$ and $\delta \to 0$. We conclude that $\log d_\infty \le -\gamma$, which is what we wanted to prove.

The double role of capacity as a conformal invariant and a geometric quantity permits us to gain relevant information about conformal mappings. For instance, if a set E is projected on a line, it is evident that the transfinite diameter decreases. Hence if the projection in any direction has length L, the capacity is $\ge L/4$. As an application, let $f(z)$ define a one-to-one conformal mapping of the unit disk, normalized by $f(0) = 0$, $|f'(0)| = 1$. Let b be a point not in the image region. Then $1/f(1/z)$ gives a normalized mapping of the unit disk on an unbounded region Ω whose complement E has capacity 1 and comprises the points 0 and $1/b$. Since E is connected it has a projection of length $\ge 1/|b|$. Hence $1 \ge \frac{1}{4}|b|$ or $|b| \ge \frac{1}{4}$. This is the famous *one-quarter theorem*.

2-4   SUBSETS OF A CIRCLE

Recall that the Dirichlet integral of a function u over a region Ω is defined by

$$D_\Omega(u) = \iint_\Omega \left[\left(\frac{\partial u}{\partial x}\right)^2 + \left(\frac{\partial u}{\partial y}\right)^2 \right] dx\, dy.$$

We shall also make use of the mixed Dirichlet integral

$$D_\Omega(u,v) = \iint_\Omega \left(\frac{\partial u}{\partial x} \frac{\partial v}{\partial x} + \frac{\partial u}{\partial y} \frac{\partial v}{\partial y} \right) dx\, dy,$$

which satisfies $D_\Omega(u,v)^2 \leq D_\Omega(u) D_\Omega(v)$. (The subscript Ω is often omitted.)

In this section we study the capacity of a closed subset E of the unit circle. It will be shown that the capacity has an extremal property which links the Dirichlet integral of a harmonic function in the unit disk Δ with its values on E and at the origin. We denote the complement of E by Ω, its Green's function by $g(z)$, and the Robin constant by $\gamma = -\log \operatorname{cap} E$.

> **Theorem 2-4**   Suppose $u(z)$ is harmonic in Δ, $u(0) = 1$, and $\limsup u(z) \leq 0$ as z approaches E. Then $D(u) \geq \pi/\gamma$ with equality for $u = g(z)/\gamma$.

PROOF   It will first be assumed that E consists of a finite number of arcs and that u is of class C^1 on the closed disk. The functions $g(z) - g(1/\bar{z})$ and $\log |z|$ are harmonic in Ω except for the same singularities at 0 and ∞, and they also have the same boundary values on E. Therefore, by the maximum principle, $g(z) - g(1/\bar{z}) = \log |z|$. This shows that $g(0) = \gamma$ and also that $\partial g/\partial r = \frac{1}{2}$ on E', the complement of E with respect to the unit circle. Since g is positive, it is furthermore true that $\partial g/\partial r < 0$ on E. There is a slight singularity at the end points of the arcs that constitute E; by standard use of the reflection principle one shows that the gradient of g is of the order of $1/\sqrt{\rho}$, where ρ is the distance from the nearest end point.

On taking all these properties into account we easily obtain

$$D_\Delta(u,g) = \int_0^{2\pi} u \frac{\partial g}{\partial r} d\theta \geq \frac{1}{2} \int_{E'} u\, d\theta \geq \frac{1}{2} \int_0^{2\pi} u\, d\theta = \pi$$

and

$$D_\Delta(g) = \int_0^{2\pi} g \frac{\partial g}{\partial r} d\theta = \frac{1}{2} \int_{E'} g\, d\theta = \frac{1}{2} \int_0^{2\pi} g\, d\theta = \pi\gamma.$$

Hence $\pi^2 \leq D_\Delta(u) D_\Delta(g) = \pi\gamma D_\Delta(u)$, and we have shown that $D_\Delta(u) \geq \pi/\gamma$. Equality occurs for $u = g/\gamma$.

An arbitrary set E can be represented as the intersection of sets E_n, each consisting of a finite number of arcs. The result we have already proved is applicable to $u_1(z) = (1 - \epsilon)^{-1}[u(rz) - \epsilon]$ and E_n for $\epsilon > 0$, $r < 1$, and sufficiently large n. Since cap $E_n \to$ cap E and $D_\Delta(u_1) \to D_\Delta(u)$, the full theorem follows.

In preparation for the next section we shall determine the capacity of an arc on the unit circle of length α. Let it be the arc between $e^{-i\alpha/2}$ and $e^{i\alpha/2}$. Its complement is mapped on the exterior of a disk centered at the origin by the function

$$f(z) = \frac{1}{2}[z - 1 + \sqrt{(z - e^{i\alpha/2})(z - e^{-i\alpha/2})}],$$

where the square root is asymptotically equal to z at ∞. Indeed, explicit computation gives

$$f(e^{i\theta}) = e^{i\theta/2}\left(i \sin \frac{\theta}{2} + \sqrt{\sin^2 \frac{\alpha}{4} - \sin^2 \frac{\theta}{2}}\right),$$

so that $|f(e^{i\theta})| = \sin (\alpha/4)$ for $|\theta| < \alpha/2$. Since $f(z)$ is normalized at ∞ the capacity of the arc is $\sin (\alpha/4)$.

2-5   SYMMETRIZATION

We would like to show that a set E on $|z| = 1$ of given length L has a minimum capacity if it consists of a single arc. In other words, we claim that cap $E \geq \sin (L/4)$. This is practically trivial if E is contained in a half circle, for then a contraction makes all distances smaller so that the transfinite diameter decreases. But if E does not lie on a half circle, this reasoning does not apply and we must look for an altogether different approach.

We shall reach our goal by combining Theorem 2-4 with a "symmetrization theorem." There are many such theorems, and we shall be concerned only by a special, but rather typical case. The kind of symmetrization we have in mind is known as *circular symmetrization*.

Let $g(\theta)$ be a measurable real-valued function defined for $0 \leq |\theta| \leq \pi$. Set $m(t)$ equal to the measure of the set on which $g(\theta) \leq t$. Note that $m(t)$ is nondecreasing and continuous on the right: $m(t + 0) = m(t)$. Two functions are said to be *equimeasurable* if they give rise to the same $m(t)$. We wish to construct a function $g^*(\theta)$ which is equimeasurable with $g(\theta)$, even, and nonincreasing for $\theta \geq 0$. The existence of g^* is quite obvious if $m(t)$ is continuous and strictly increasing, for then we merely set $g^*(\theta) = g^*(-\theta) = m^{-1}(2\theta)$, where m^{-1} is the uniquely defined inverse of m. In the general case we define $g^*(\theta)$ as the least upper bound of all t with $m(t - 0) \leq 2\theta$. We must prove, carefully, that g^* and g are equimeasurable.

It is clear that g^* is nondecreasing. Therefore, the inequality $g^*(\theta) \leq t_0$ holds either in an open interval $|\theta| < \theta_0$, or on a closed interval $|\theta| \leq \theta_0$. We have to show that $m(t_0) = 2\theta_0$.

> *1* If $m(t_0) > 2\theta_0$, then $m(t_0) > 2\theta_0 + 2\epsilon$ for some $\epsilon > 0$, and hence $m(t_0 + \delta - 0) > 2\theta_0 + 2\epsilon$ for all $\delta > 0$. By the definition of g^* this implies $g^*(\theta_0 + \epsilon) \leq t_0 + \delta$, and hence $g^*(\theta_0 + \epsilon) \leq t_0$. This contradicts the definition of θ_0.
>
> *2* If $m(t_0) < 2\theta_0$, we can find ϵ and $\delta > 0$ such that $m(t_0 + \delta - 0) < 2\theta_0 - 2\epsilon$. This implies $g^*(\theta_0 - \epsilon) \geq t_0 + \delta$, which again contradicts the definition of θ_0. The only remaining possibility is to have $m(t_0) = 2\theta_0$.

For the sake of simplicity the Poisson integral in the unit disk formed with the boundary values $g(\theta)$ will be denoted by $g(z)$. We assume of course that $g(\theta)$ is integrable. The notation $D_\Delta(g)$ will be simplified to $D(g)$. We shall need a formula that expresses $D(g)$ in terms of the boundary values $g(\theta)$.

Theorem 2-5

$$D(g) = \frac{1}{8\pi} \int_0^{2\pi} \int_0^{2\pi} \left(\frac{g(\theta) - g(\theta')}{\sin\left[(\theta - \theta')/2\right]} \right)^2 d\theta\, d\theta'.$$

PROOF   We complete g to an analytic function $f = g + ih$ normalized by $h(0) = 0$. The Dirichlet integral extended over $\Delta_r = \{|z| < r\}$ can be expressed by

$$D_r(g) = -\frac{i}{2} \int_{|z| = r} \bar{f} f'\, dz.$$

In this formula we have to substitute

$$f(z) = \frac{1}{2\pi} \int_0^{2\pi} \frac{e^{i\theta} + z}{e^{i\theta} - z} g(\theta)\, d\theta$$

$$f'(z) = \frac{1}{\pi} \int_0^{2\pi} \frac{e^{i\theta}}{(e^{i\theta} - z)^2} g(\theta)\, d\theta.$$

A simple residue computation gives

$$\int_{|z|=r} \frac{e^{-i\theta} + \bar{z}}{e^{-i\theta} - \bar{z}} \frac{e^{i\theta'}}{(e^{i\theta'} - z)^2} \, dz = \int_{|z|=r} \frac{(z + r^2 e^{i\theta}) e^{i\theta'}}{(z - r^2 e^{i\theta})(e^{i\theta'} - z)^2} \, dz$$

$$= 2\pi i \frac{2r^2 e^{i(\theta+\theta')}}{(e^{i\theta'} - r^2 e^{i\theta})^2} = 4\pi i r^2 (e^{i(\theta'-\theta)/2} - r^2 e^{i(\theta-\theta')/2})^{-2},$$

and we obtain

$$D_r(g) = \frac{r^2}{\pi} \int_0^{2\pi} \int_0^{2\pi} (e^{i(\theta'-\theta)/2} - r^2 e^{i(\theta-\theta')/2})^{-2} g(\theta) g(\theta') \, d\theta \, d\theta'. \quad (2\text{-}8)$$

As a special case of (2-8),

$$\int_0^{2\pi} \int_0^{2\pi} (e^{i(\theta'-\theta)/2} - r^2 e^{i(\theta-\theta')/2})^{-2} \, d\theta \, d\theta' = 0,$$

and since

$$\int_0^{2\pi} (e^{i(\theta'-\theta)/2} - r^2 e^{i(\theta-\theta')/2})^{-2} \, d\theta \quad (2\text{-}9)$$

is obviously independent of θ', the integral (2-9) is identically zero. It follows that

$$\int_0^{2\pi} \int_0^{2\pi} (e^{i(\theta'-\theta)/2} - r^2 e^{i(\theta-\theta')/2})^{-2} g(\theta)^2 \, d\theta \, d\theta' = 0,$$

and (2-8) can thus be replaced by

$$D_r(g) = -\frac{r^2}{2\pi} \int_0^{2\pi} \int_0^{2\pi} \left[\frac{g(\theta) - g(\theta')}{e^{i(\theta'-\theta)/2} - r^2 e^{i(\theta-\theta')/2}} \right]^2 d\theta \, d\theta'. \quad (2\text{-}10)$$

As $r \to 1$ we expect the right-hand member of (2-10) to converge to

$$I(g) = \frac{1}{8\pi} \int_0^{2\pi} \int_0^{2\pi} \left(\frac{g(\theta) - g(\theta')}{\sin[(\theta - \theta')/2]} \right)^2 d\theta \, d\theta'.$$

If $I(g) < \infty$, it is easy to prove that this is indeed so, for the identity

$$|e^{i(\theta'-\theta)/2} - r^2 e^{i(\theta-\theta')/2}|^2 = (1 - r^2)^2 + 4r^2 \sin^2 \frac{\theta - \theta'}{2} \quad (2\text{-}11)$$

shows that $I(g)$ is a convergent majorant of $D_r(g)$ so that Lebesgue's theorem of dominated convergence is applicable.

It remains to show that $I(g)$ converges when $D(g) < \infty$. First, if the Dirichlet integral is finite, the radial limit $g(\theta) = \lim_{r \to 1} g(re^{i\theta})$ exists almost everywhere and is square integrable. This is an immediate consequence of the Schwarz inequality, for instance, in the form

$$\int_0^{2\pi} \left[\int_{\frac{1}{2}}^1 \frac{\partial}{\partial r} g(re^{i\theta}) \, dr \right]^2 d\theta \leq \frac{1}{2} \int_{\frac{1}{2}}^1 \int_0^{2\pi} \left(\frac{\partial g}{\partial r} \right)^2 dr \, d\theta < D(g) < \infty.$$

Furthermore, it is well known that the Poisson integral formed with the values $g(\theta)$ is equal to $g(z)$. Now we apply the theorem to $g_r(\theta) = g(re^{i\theta})$. We know that $I(g_r) = D(g_r) = D_r(g)$ and that $g_r(\theta) \to g(\theta)$ almost everywhere. It follows by Fatou's lemma that $I(g) \leq \lim I(g_r) = D(g) < \infty$. The theorem is proved.

We return to the symmetrized function g^*.

Theorem 2-6 $D(g) \geq D(g^*)$.

PROOF   We introduce the auxiliary integral

$$E_r(g) = \frac{r^2}{2\pi} \int_0^{2\pi} \int_0^{2\pi} \frac{[g(\theta) - g(\theta')]^2}{(1 - r^2)^2 + 4r^2 \sin^2[(\theta - \theta')/2]}\, d\theta\, d\theta'.$$

By (2-9), $D_r(g) \leq E_r(g) \leq I(g)$. If we can show that $E_r(g^*) \leq E_r(g)$, the theorem will follow, for then $D_r(g^*) \leq E_r(g^*) \leq E_r(g) \leq I(g)$, and hence in the limit $D(g^*) = I(g^*) \leq I(g)$.

The missing part will be supplied by a more general symmetrization lemma.

Lemma 2-2   Let $u(\theta)$ and $v(\theta)$ be measurable and nonnegative for $-\pi \leq \theta \leq \pi$ and suppose that $K(t)$ is nonnegative and nonincreasing for $0 \leq t \leq 1$. Then the integral

$$J(u,v) = \int_{-\pi}^{\pi} \int_{-\pi}^{\pi} u(\theta)v(\theta')K\left(\sin\frac{|\theta - \theta'|}{2}\right) d\theta\, d\theta'$$

satisfies $J(u,v) \leq J(u^*,v^*)$.

The application to the proof of $E_r(g^*) \leq E_r(g)$ is rather obvious. First, we may assume g to be bounded, and since added constants have no influence, we may take g positive. If we develop the square in $E_r(g)$, it is obvious that the terms containing $g(\theta)^2$ and $g(\theta')^2$ do not change when g is replaced by g^*. It is therefore sufficient to show that the term with $g(\theta)g(\theta')$ increases, and that follows from the lemma since the kernel is a decreasing function of $\sin(|\theta - \theta'|/2)$.

PROOF   Define $u_+(\theta) = \max[u(\theta),u(-\theta)]$ if $\theta \in (0,\pi)$ and $u_+(\theta) = \min[u(\theta),u(-\theta)]$ if $\theta \in (-\pi,0)$. In other words, if we think of $u(\theta)$ as being defined on the unit circle, we change the values at conjugate points in such a way that the larger value is taken on the upper half circle. With a similar change of v we claim that $J(u_+,v_+) \geq J(u,v)$.

For fixed $\theta,\theta' \in (0,\pi)$ we have to compare

$$\Delta(u,v) = [u(\theta)v(\theta') + u(-\theta)v(-\theta')]K\left(\sin\frac{|\theta - \theta'|}{2}\right)$$
$$+ [u(\theta)v(-\theta') + u(-\theta)v(\theta')]K\left(\sin\frac{\theta + \theta'}{2}\right)$$

with $\Delta(u_+,v_+)$. In view of obvious symmetry considerations it is sufficient to consider the case in which $u_+(\theta) = u(\theta)$ and $v_+(\theta) = v(-\theta)$. It is seen that

$$\Delta(u_+,v_+) - \Delta(u,v)$$
$$= [u(\theta) - u(-\theta)][v(-\theta') - v(\theta')]\left[K\left(\sin\frac{|\theta - \theta'|}{2}\right) - K\left(\sin\frac{\theta + \theta'}{2}\right)\right],$$

a product of three positive factors. Hence $J(u_+,v_+) \geq J(u,v)$.

In proving the lemma we may assume that u, v, and K are bounded. Next, we can replace u and v by step functions that assume only a finite number of values, each on a semiopen interval. We set

$$u = a_1u_1 + \cdots + a_nu_n,$$

where the u_i are characteristic functions of intervals and $a_i > 0$. It is evident that $u^* = a_1u_1^* + \cdots + a_nu_n^*$, and it is similarly evident that

$$v = b_1v_1 + \cdots + b_mv_m.$$

The lemma needs to be proved only for u_i and v_j.

Accordingly, let u and v be the characteristic functions of two arcs α and β on the unit circle. It is evident that $J(u,v)$ is invariant under rotations of the circle. Therefore we can assume that the midpoints of α and β are symmetric with respect to the real axis, for instance, with α having its midpoint on the upper half of the circle. But in this situation $u_+(\theta) = u(\theta)$ and $v_+(\theta') = v(-\theta')$. Thus u_+ and v_+ are characteristic functions of arcs with a common midpoint. This implies $J(u^*,v^*) = J(u_+,v_+)$, and we have already proved that $J(u_+,v_+) \geq J(u,v)$. The lemma is proved.

The application to capacity is quite simple.

Theorem 2-7 A closed set E on the unit circle with length α has a capacity at least equal to $\sin(\alpha/4)$.

PROOF Let $g(z)$ be the Green's function of the complement of E, and γ the Robin constant, so that cap $E = e^{-\gamma}$. We denote by $g^*(z)$

the harmonic function with the symmetrized equimeasurable boundary values $g^*(\theta)$. It is equal to zero on an arc E^* of length α, and we set $\text{cap } E^* = e^{-\gamma^*} = \sin(\alpha/4)$. Because g and g^* are equimeasurable on the unit circle, they have the same value at the origin, and in the course of the proof of Theorem 2-4 it was shown that this value is γ. Now apply Theorem 2-4 to g/γ and g^*/γ. It yields $D(g/\gamma) = \pi/\gamma$ and $D(g^*/\gamma) \geq \pi/\gamma^*$. Since $D(g) \geq D(g^*)$, it follows that $\gamma \leq \gamma^*$, $\text{cap } E \geq \text{cap } E^*$.

NOTES   The transfinite diameter was introduced by Fekete [19]. Important simplifications were given by Szegö [59] and in Polya and Szegö [54].

In spite of its classic origin potential theory suffered a long time from lack of rigor and did not come into its own until the appearance of Frostman's thesis [20] in 1935. More work on the foundations was done by H. Cartan and in many articles by Brelot. For later developments the reader is referred to the excellent survey and bibliography by Brelot [9].

The representation of the Dirichlet integral in terms of the boundary values (Theorem 2-5) is due to Douglas [18]. Symmetrization techniques were used extensively by Polya. All rearrangement theorems, of which Lemma 2-2 is a sample, are based on a fundamental idea of Hardy and Littlewood [26]. Theorem 2-7 was probably first proved, but not published, by Beurling.

EXERCISES

1 Show that the transfinite diameter of a set is equal to that of its boundary.

2 Let E be a closed, bounded, connected set, and let Ω denote the unbounded component of its complement. By Riemann's mapping theorem there is a normalized conformal mapping

$$f(z) = z + a_0 + a_1 z^{-1} + \cdots$$

of Ω on the outside of a circular disk whose radius R is known as the outer conformal radius of Ω. Show that $R = \text{cap } E$.

3 (Hayman) Let $F(w) = \alpha w + b_0 + b_1 w^{-1} + \cdots$ be analytic in $1 < |w| < \infty$. Denote by Q the set of all values that $F(w)$ does not assume. Then $\text{cap } Q \leq \alpha$ with equality only if F is one to one.

4 (Pommerenke) Study the mapping of $|z| > 1$ by $(z^n + 2 + z^{-n})^{1/n}$ and use it to find the capacity of a star formed by n equally spaced segments of unit length issuing from a point.

HARMONIC MEASURE

3-1 THE MAJORIZATION PRINCIPLE

Harmonic functions satisfy the maximum principle. Elementary but systematic use of the maximum principle leads to important methods for majorizing and minorizing harmonic and analytic functions. In this connection the emphasis is never on great generality but on usefulness. For this reason we shall deal only with situations in which the existence theorems needed are virtually trivial.

It has been found extremely useful to introduce the notion of *harmonic measure*. In its most primitive form it arises as follows.

Let Ω be a region in the extended complex plane whose boundary $\partial\Omega$ consists of a finite number of disjoint Jordan curves. Suppose that the boundary $\partial\Omega$ is divided into two parts E and E', each consisting of a finite number of arcs and closed curves; it is immaterial whether the end points of the arcs are included or not. There exists a unique bounded harmonic function $\omega(z)$ in Ω such that $\omega(z) \to 1$ when z tends to an interior point of E and $\omega(z) \to 0$ when z tends to an interior point of E'. The values of ω lie strictly between 0 and 1. The number $\omega(z)$ is called the harmonic measure of E at the point z with respect to the region Ω. It is also denoted by $\omega(z,\Omega,E)$.

We remind the reader that the uniqueness depends essentially on the boundedness of ω. It follows from the Lindelöf form of the maximum principle, which we state and prove for the benefit of the reader.

Lindelöf's maximum principle   Let $u(z)$ be harmonic and bounded above, $u(z) \leq M$, in a region Ω whose boundary is not a finite set. Suppose that $\lim \sup_{z \to \zeta} u(z) \leq m$ for all boundary points ζ with a finite number of exceptions. Then $u(z) \leq m$ in Ω.

PROOF   Assume first that Ω is bounded and denote its diameter by d, the exceptional points by ζ_j. In this case it suffices to apply the ordinary maximum principle to

$$u(z) + \epsilon \sum_j \log \frac{|z - \zeta_j|}{d}$$

with $\epsilon > 0$, and then let ϵ tend to zero. With the help of an inversion the same reasoning can be applied as soon as Ω has an exterior point. In the absence of exterior points we choose a positive number R different from all $|\zeta_j|$. Let Ω_1 and Ω_2 be the parts of Ω with $|z| < R$ and $|z| > R$, respectively. If $u(z) \leq m$ on $\Omega \cap \{|z| = R\}$, we can apply the result separately in Ω_1 and Ω_2. If not, u will have a maximum $> m$ on $\Omega \cap \{|z| = R\}$, and this is a maximum for all of Ω. But then u would be a constant $> m$ and could not satisfy the boundary condition.

EXAMPLE 3-1   Let Ω be the upper half plane and E a finite union of segments of the real axis. Then $\omega(z,\Omega,E)$ is $1/\pi$ times the total angle under which E is seen from the point z.

EXAMPLE 3-2   Let Ω be a circular disk and E an arc of the circle with central angle α. The harmonic measure is $\omega(z) = (2\theta - \alpha)/2\pi$ where θ is the angle subtended by E at z.

EXAMPLE 3-3   Let Ω be an annulus $r_1 < |z| < r_2$ and E the outer circle $|z| = r_2$. The harmonic measure at z is $\log(|z|/r_1) : \log(r_2/r_1)$.

For some purposes it is preferable to consider a slightly more general situation. Let Ω be an open set and α a closed set in the extended plane. We denote by E the part of the boundary of $\Omega - \alpha$ that belongs to α. It will be assumed that the geometric situation is so simple that it is possible to form $\omega(z,\Omega - \alpha,E)$, separately for each component of $\Omega - \alpha$. This is called the harmonic measure of α with respect to Ω, and the notation is simplified to $\omega(z,\Omega,\alpha)$. Roughly speaking, ω is harmonic and bounded in $\Omega - \alpha$, equal to 1 on α, and zero on the rest of the boundary.

The principle of majorization is stated as follows:

Theorem 3-1 Given two pairs (Ω,α) and (Ω^*,α^*), let f be analytic in $\Omega - \alpha$ with values in Ω^*, and such that $z \to \alpha$ implies $f(z) \to \alpha^*$. Under these circumstances $\omega(z,\Omega,\alpha) \leq \omega(f(z),\Omega^*,\alpha^*)$ in $f^{-1}(\Omega^* - \alpha^*)$.

We write $\omega = \omega(z,\Omega,\alpha)$, $\omega^* = \omega(f(z),\Omega^*,\alpha^*)$ and apply the maximum principle to $\omega - \omega^*$ in a component of $f^{-1}(\Omega^* - \alpha^*)$. As z approaches the boundary of this component, either z tends to a boundary point of Ω which is not on α, or $f(z)$ tends to α^*. In either case $\limsup (\omega - \omega^*) \leq 0$, except when $f(z)$ tends to an end point of the boundary arcs of $\Omega - \alpha^*$ that lie on α^*. If there are only finitely many such points, the maximum principle remains valid, and we conclude that $\omega \leq \omega^*$ throughout $f^{-1}(\Omega^* - \alpha^*)$.

Corollary $\omega(z,\Omega,\alpha)$ increases with Ω and α.

This is the theorem applied to the identity mapping.

As another application let Ω^* be the disk $|w| < M$ and α^* a closed disk $|w| \leq m < M$. Then

$$\omega(f(z),\Omega^*,\alpha^*) = \frac{\log [M/|f(z)|]}{\log M/m}$$

and we obtain the "two-constant theorem," Theorem 3-2.

Theorem 3-2 If $|f(z)| \leq M$ in Ω and $|f(z)| \leq m$ on α, then $|f(z)| \leq m^\theta M^{1-\theta}$ in the part where $\omega(z,\Omega,\alpha) \geq \theta$.

When both regions are annuli we obtain as a special case Hadamard's three-circle theorem which is best formulated as a determinant inequality

$$\begin{vmatrix} 1 & 1 & 1 \\ \log r & \log \rho & \log R \\ \log M(r) & \log M(\rho) & \log M(R) \end{vmatrix} \geq 0 \qquad (3\text{-}1)$$

for $r < \rho < R$. Here $M(r)$ is the maximum of $|f(z)|$ for $|z| = r$. The inequality shows that $\log M(r)$ is a convex function of $\log r$. Needless to say the same is true of the maximum of any subharmonic function.

3-2 APPLICATIONS IN A HALF PLANE

The next three theorems are typical applications of the majorization principle.

Theorem 3-3 Let $f(z)$ be analytic and bounded for $y > 0$, continuous on the real axis. If $f(x) \to c$ for $x \to +\infty$, then $f(z) \to c$ in any angle $0 \leq \arg z \leq \pi - \delta$, $\delta > 0$.

PROOF We may assume that $|f(z) - c| < 1$ in the half plane, and that $|f(x) - c| < \epsilon$ for $x > x_0$. The two-constant theorem gives $|f(z) - c| < \epsilon^{\delta/2\pi}$ when $\arg (z - x_0) < \pi - \delta/2$. In particular, this inequality holds as soon as $|z| > x_0$, $\arg z < \pi - \delta$.

Theorem 3-4 (Phragmén-Lindelöf's principle) Let $f(z)$ be analytic for $y > 0$, continuous with $|f(x)| \leq 1$ for all real x. Then either $|f(z)| \leq 1$ in the entire half plane, or the maximum modulus $M(r)$ on $|z| = r$ satisfies $\liminf_{r\to\infty} r^{-1} \log M(r) > 0$.

PROOF In Fig. 3-1 the harmonic measure of the half circle is $2\theta/\pi$. Hence $|f(z)| \leq M(R)^{2\theta/\pi}$. But $R\theta$ tends to a finite limit when z is fixed and $R \to \infty$. The theorem follows.

Theorem 3-5 (Lindelöf) Let $f(z)$ be analytic and bounded in a half plane and assume that $f(z) \to c$ along an arc γ tending to ∞. Then $f(z) \to c$ uniformly in any interior angle.

PROOF We may assume that $c = 0$ and $|f(z)| \leq 1$. Replace γ by its subarc from the last intersection with $|z| = R$ (Fig. 3-2). It divides $\{|z| > R\}$ into two regions Ω' and Ω'', and we may suppose that $|f(z)| < \epsilon$

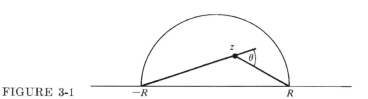

FIGURE 3-1

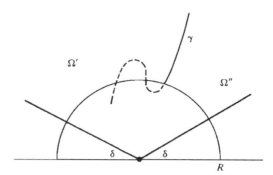

FIGURE 3-2

on γ. The harmonic measure $\omega(z,\Omega',\gamma)$ is greater than or equal to the harmonic measure of $[R, +\infty]$ with respect to $\{|z| > R\}$. If the latter is denoted by $\omega(z,R)$, it is clear that $\omega(z,R) = \omega(z/R,1)$. But $\omega(z,1)$ has a positive minimum λ, for instance, on the arc $\{|z| = 2,\ \arg z \leq \pi - \delta\}$. We conclude that $|f(z)| \leq \epsilon^\lambda$ when $|z| = 2R$ and at the same time $z \in \Omega'$, $\arg z \leq \pi - \delta$. The same estimate holds when $z \in \Omega''$, $\arg z \geq \delta$. The theorem is proved.

Corollary A bounded analytic function in the half plane cannot tend to different limits along two paths leading to infinity.

A limit along a path that leads to infinity is known as an asymptotic value. The corollary asserts that a bounded analytic function cannot have two different asymptotic values in a half plane. The same is true when the half plane is replaced by an angle.

3-3 MILLOUX'S PROBLEM

Suppose that $f(z)$ is analytic and $|f(z)| \leq M$ in the unit disk Δ. Assume further that the minimum of $|f(z)|$ on every circle $\{|z| = r\}, 0 < r < 1$, is $\leq m < M$. How large can $|f(z_0)|$ be at a given point? This is a natural setting for harmonic measure. Indeed, we know that $|f(z)| \leq m$ on a certain set α and $|f(z)| \leq M$ elsewhere. Hence by the two-constant theorem, $|f(z_0)| \leq m^\omega M^{1-\omega}$, where ω is the harmonic measure of α with respect to Δ at the point z_0. What we need therefore is a lower bound for ω which does not depend on the shape of α but merely on the fact that α intersects all the concentric circles.

This problem was proposed by Milloux [38] and aroused considerable interest. It was solved independently by R. Nevanlinna [44] and

Beurling [61]. We shall give an account of the development that led to the solution.

Because of the rotational symmetry we may assume that z_0 is positive, and we set $z_0 = r_0 < 1$. It is to be expected that ω is a minimum when α is the negative radius, and we shall ultimately show that this is indeed the case. For the negative radius one easily finds that $\omega_0 = \omega(r_0)$ is given by

$$1 - \omega_0 = \frac{4}{\pi} \text{ arc tan } \sqrt{r_0}.$$

If we show that $\omega(z_0,\Delta,\alpha) \geq \omega_0$, it will follow that $|f(z_0)| \leq m^{\omega_0} M^{1-\omega_0}$ not only for the functions f originally under consideration, but also for functions f that are analytic only in $\Delta - \alpha$ and satisfy lim sup $|f(z)| \leq m$ as $z \to \alpha$. Indeed, since the maximum principle is being applied to log $|f(z)|$, it is not even necessary that $f(z)$ be single-valued as long as it has a single-valued modulus. Thus the solution applies to a larger class, and it will give the best possible estimate for the larger class; but it does not give the best answer to the original question. Although the answer is known (Heins [30]), we shall be content to solve the easier problem of finding the best lower bound for $\omega(z_0)$.

Even before Milloux an important special case was treated by Carleman [12]. He assumed that the region $\Delta - \alpha$ is simply connected. Under this hypothesis we can map $\Delta - \alpha$ by a branch of log z. The image region lies in the left half plane, and it has the property of intersecting each vertical line along segments whose total length is at most 2π. By conformal invariance $1 - \omega(r_0)$ is the harmonic measure at log r_0 of that part of the boundary of the image which lies on the imaginary axis. It is majorized by the harmonic measure with respect to the full left half plane. The angle subtended at log r_0 by segments of given total length is a maximum for a single segment, symmetrically placed with respect to the real axis. It follows that

$$1 - \omega(r_0) \leq \frac{2}{\pi} \text{ arc tan } \frac{\pi}{\log (1/r_0)}. \tag{3-2}$$

This is a rather poor estimate, the reason being that the half plane is the image of a Riemann surface with infinitely many sheets rather than the image of a plane region. Carleman used the following idea to improve the result. Assume that $|f| \leq 1$ on α and let $M(r)$ denote the maximum modulus of f on $|z| = r$. We use (3-2) in a disk of radius $r > r_0$ to obtain

$$\frac{\log M(r_0)}{\log M(r)} \leq \frac{2}{\pi} \text{ arc tan } \frac{\pi}{\log (r/r_0)}. \tag{3-3}$$

On letting r_0 tend to r this becomes a differential inequality

$$\frac{d \log M(r)}{d \log r} \geq \frac{2}{\pi^2} \log M(r),$$

and integration from r_0 to 1 gives

$$\log \log M(1) - \log \log M(r_0) \geq \frac{2}{\pi^2} \log \frac{1}{r_0}. \qquad (3\text{-}4)$$

[We have tacitly assumed $M(r)$ to be differentiable. This can be avoided by passing directly from (3-3) to (3-4).]

We apply (3-4) with $\log |f| = 1 - \omega$ and obtain the estimate

$$1 - \omega(r_0) \leq r_0^{2/\pi^2},$$

which is an improvement over (3-2).

More important, it is possible to take into account the "thickness" of the set α. For this purpose let $\theta(r)$ denote the total angle of the arcs on $|z| = r$ complementary to α. It is rather obvious that (3-3) can be replaced by

$$\frac{d \log M(r)}{d \log r} \leq \frac{4}{\pi \theta(r)} \log M(r),$$

from which it follows that

$$\log \log M(1) - \log \log M(r_0) \geq \frac{4}{\pi} \int_{r_0}^{1} \frac{dr}{r\theta(r)},$$

and finally

$$\log [1 - \omega(r_0)] \leq \exp \left[-\frac{4}{\pi} \int_{r_0}^{1} \frac{dr}{r\theta(r)} \right].$$

Although this is not precise, it is at any rate a good estimate obtained by very elementary methods.

We pass now to Beurling's solution of the problem, or rather of a more general problem. Let α be a closed set in Δ, and let α^* be its circular projection on the negative radius. In other words α^* consists of all $-r$ such that there exists a $z \in \alpha$ with $|z| = r$.

Theorem 3-6 (**Beurling**) $\omega(r_0,\Delta,\alpha) \geq \omega(r_0,\Delta,\alpha^*)$.

The proof can be given for arbitrary α (with an appropriate definition of ω), but we shall be content to consider the case in which α^* consists of a finite number of segments. In this case the existence of

$\omega^*(z) = \omega(z,\Delta,\alpha^*)$ is clear, and the existence of $\omega(z,\Delta,\alpha)$ will be taken for granted as part of the hypothesis.

We know that

$$g(z,\zeta) = \log\left|\frac{1 - \bar{z}\zeta}{z - \zeta}\right|$$

is the Green's function of Δ with a pole at ζ, and we note the inequalities

$$g(|z|, -|\zeta|) \le g(z,\zeta) \le g(|z|,|\zeta|) = g(-|z|, -|\zeta|). \tag{3-5}$$

Green's formula gives the representation

$$\omega^*(z) = -\frac{1}{\pi} \int_{\alpha^*} g(\zeta,z) \frac{\partial \omega^*(\zeta)}{\partial n} |d\zeta|, \tag{3-6}$$

where we have added the two equal normal derivatives, both in the direction of the inner normal, and hence negative.

In (3-6) we replace each $\zeta \in \alpha^*$ by a $\zeta' \in \alpha$ such that $|\zeta'| = |\zeta|$. This can be done in such a way that $g(\zeta',z)$ is a measurable function. We can therefore form the function

$$u(z) = -\frac{1}{\pi} \int_{\alpha^*} g(\zeta',z) \frac{\partial \omega^*(\zeta)}{\partial n} |d\zeta|.$$

It is obvious that u is harmonic in $\Delta - \alpha$ and that $u = 0$ on $|z| = 1$. With the help of (3-5) we see that $u(z) \le \omega^*(-|z|) \le 1$ for all $z \in \Delta - \alpha$, and it follows by the maximum principle that $u(z) \le \omega(z)$. On the other hand $u(r_0) \ge \omega^*(r_0)$, again by (3-5), and we have proved that $\omega(r_0) \ge \omega^*(r_0)$.

3-4 THE PRECISE FORM OF HADAMARD'S THEOREM

We return to Hadamard's three-circle theorem which we derived as a consequence of Theorem 3-2, the two-constant theorem. We shall assume that $f(z)$ is analytic for $1 \le |z| \le R$ and that $|f| \le 1$ on $|z| = 1$, $|f| \le M$ on $|z| = R$; we assume $M > 1$. The inequality (3-1) takes the form

$$\log M(r) \le \frac{\log R}{\log r} \log M.$$

Equality can hold only if

$$\log |f(z)| = \frac{\log |z|}{\log R} \log M,$$

and we then have

$$\int_{|z|=r} \frac{\partial \log |f|}{\partial r}\, r\, d\theta = 2\pi \frac{\log M}{\log R}.$$

But if $f(z)$ is single-valued, as we assume, the left-hand side is an integral multiple of 2π, and hence equality can occur only if M is an integral power of R. When this is not the case, there must be a smaller bound for $M(r)$, and it becomes an interesting problem to find the best bound for arbitrary r, R, and M.

This problem was solved by Teichmüller [60], and independently by Heins [29]. Teichmüller's solution is more explicit, and it is the one we shall present.

We choose a fixed ρ, $1 < \rho < R$, and denote the Green's function of the annulus $1 < |z| < R$ with a pole at $-\rho$ by $g(z) = g(z, -\rho)$. The harmonic measure of the outer circle is $\omega(z) = \log |z|/\log R$.

Lemma 3-1   The maximum of $g(z)$ on $|z| = r$ is attained at $-r$, the minimum at r. As a consequence the radial derivative $\partial g/\partial r$ has its minimum on $|z| = R$ at R, and its maximum on $|z| = 1$ at 1. Moreover, $g(r)$ is a strictly concave function of $\log r$, that is, $rg'(r)$ is strictly decreasing for $1 < r < R$.

PROOF   We write $z = re^{i\theta}$ and consider the harmonic function $\partial g/\partial \theta$. It vanishes on $|z| = 1$ and $|z| = R$, and it is also zero on the real axis because of the symmetry. Near the pole it behaves as

$$\operatorname{Im}\left[z \frac{d}{dz} \log (z + \rho) \right] = \operatorname{Im}\left[\frac{z}{(z + \rho)} \right]$$

which is positive in the upper half plane. Hence $\partial g/\partial \theta > 0$ in the upper half of the annulus, which proves the assertion about the maximum and minimum. The statements about the maximum and minimum of $\partial g/\partial r$ are a trivial consequence.

The three-circle theorem for harmonic functions states that the maximum of a harmonic function on $|z| = r$ is a convex function of $\log r$, and strictly convex except for functions of the form $a \log |z| + b$. It follows that the minimum $g(r)$ is concave in the intervals $(1,\rho)$ and (ρ, R) and hence $rg'(r)$ is decreasing in $(1, R)$. If $g(r)$ were equal to $a \log r + b$ in an interval, we would have $g(z) = a \log |z| + b$ in the corresponding annulus. This would imply that the analytic function $r\, \partial g/\partial r - i\, \partial g/\partial \theta$ reduces to a constant, which is impossible in view of the singularity at $-\rho$.

Theorem 3-7   Suppose that $f(z)$ is analytic in $1 \leq |z| \leq R$ and that $|f(z)| \leq 1$ for $|z| = 1$, $|f(z)| \leq M$ for $|z| = R$. Let m be the integer determined by $R^{m-1} < M \leq R^m$. Then

$$\log M(\rho) \leq \frac{\log M \log \rho}{\log R} - g\left(\frac{R^m}{M}\right), \tag{3-7}$$

with equality only if

$$\log |f(z)| = \omega(z) \log M - g\left(z, \frac{R^m}{M}\right). \tag{3-8}$$

PROOF   We may assume that the maximum modulus $M(\rho)$ is attained at $-\rho$. Denote the zeros of f by $a_1, \ldots, a_N$ and write C for the full boundary of the annulus. Green's formula gives

$$\log |f(-\rho)| = -\frac{1}{2\pi} \int_C \log |f| \frac{\partial g}{\partial n} |dz| - \sum_1^N g(a_i) \tag{3-9}$$

and

$$\frac{1}{2\pi} \int_C \log |f| \frac{\partial \omega}{\partial n} |dz| - \sum_1^N \omega(a_i) = \frac{1}{2\pi} \int_{|z|=R} \frac{\partial \log |f|}{\partial n} |dz|.$$

Observe that f is single-valued if and only if the right-hand member of the last equation is an integer. Our problem is to maximize (3-9) under the diophantine condition

$$\frac{1}{2\pi} \int_C \log |f| \frac{\partial \omega}{\partial n} |dz| - \sum_1^N \omega(a_i) \equiv 0 \text{ (mod 1)}. \tag{3-10}$$

We note first that $g(a_i) \geq g(|a_i|)$, by Lemma 3-1, while

$$\omega(a_i) = \omega(|a_i|).$$

For this reason we may assume that the a_i are positive. Let us write $\log |f| = -u_1$ on $|z| = 1$ and $\log |f| = \log M - u_2$ on $|z| = R$; then u_1 and u_2 are ≥ 0. The normal derivative $\partial g / \partial n$ (outer normal) is $\leq -g'(1)$ on $|z| = 1$ and $\leq g'(R)$ on $|z| = R$.

With these estimates, and by use of the relation

$$-\frac{1}{2\pi} \int_{|z|=R} \frac{\partial g}{\partial n} |dz| = \omega(-\rho) = \frac{\log \rho}{\log R},$$

it follows from (3-9) that

$$\log |f(-\rho)| \leq \frac{\log M \log \rho}{\log R} - \frac{g'(1)}{2\pi} \int_{|z|=1} u_1 \, d\theta$$

$$+ \frac{Rg'(R)}{2\pi} \int_{|z|=R} u_2 \, d\theta - \sum_1^N g(a_i), \tag{3-11}$$

with equality only if u_1 and u_2 are identically zero. For further simplification we set $g(r) = h[\omega(r)]$, $\omega(a_i) = t_i$, and

$$A_1 = (2\pi \log R)^{-1} \int_{|z|=1} u_1 \, d\theta, \qquad A_2 = (2\pi \log R)^{-1} \int_{|z|=R} u_2 \, d\theta.$$

With this notation (3-11) becomes

$$\log |f(-\rho)| \leq \frac{\log M \log \rho}{\log R} - A_1 h'(0) + A_2 h'(1) - \sum_1^N h(t_i), \quad (3\text{-}12)$$

while (3-10) can be written

$$A_1 - A_2 + \sum_1^N t_i = -\frac{\log M}{\log R} \pmod 1. \tag{3-13}$$

We shall make use of the following lemma.

Lemma 3-2 Let $h(t)$ be periodic with period 1, $h(0) = h(1) = 0$, and $h''(t) < 0$ for $0 < t < 1$. Then h is subadditive: $h(x + y) \leq h(x) + h(y)$, and more generally $h(A_1 - A_2 + t_1 + \cdots + t_N) \leq A_1 h'(0) - A_2 h'(1) + h(t_1) + \cdots + h(t_N)$ if $A_1, A_2 \geq 0$. Equality holds only if $A_1 = A_2 = 0$ and at most one t_i is not an integer.

The lemma is applicable to the function h in (3-12), extended by periodicity. Set $\beta = \omega(R^m/M) = m - (\log M/\log R)$ so that

$$h(\beta) = g\left(\frac{R^m}{M}\right),$$

m being the integer in the statement of the theorem. We obtain from (3-13) and Lemma 3-2

$$h(\beta) = h\left(A_1 - A_2 + \sum_1^N t_i\right) \leq A_1 h'(0) - A_2 h'(1) + \sum_1^N h(t_i),$$

and hence by (3-12)

$$\log |f(-\rho)| \leq \frac{\log M \log \rho}{\log R} - g\left(\frac{R^m}{M}\right),$$

which is (3-7).

For equality we must have equality in (3-12) and in the lemma. This occurs only if $A_1 = A_2 = 0$ and f has a single zero a_1. We then have

$$\log |f(z)| = \omega(z) \log M - g(z, a_1),$$

and by (3-13) $t_1 = \omega(a_1) \equiv -\log M / \log R$ (mod 1), which is possible only for $a_1 = R^m / M$. We have shown that the function f defined by (3-8) is single-valued, and that its maximum modulus $M(\rho)$ is the largest possible for all ρ.

PROOF OF LEMMA   For the first part of the lemma we may assume that $0 < x < 1$, $0 < y < 1$. If $x + y \le 1$, we have $h'(t + y) < h'(t)$ for $0 < t < x$, and hence $h(x + y) - h(y) < h(x)$. If $x + y > 1$, we have $h'(t + y - 1) > h'(t)$ for $x < t < 1$, and hence $h(y) - h(x + y - 1) > -h(x)$, so that in both cases $h(x + y) < h(x) + h(y)$.

Repeated use gives $h(t_1 + \cdots + t_N) \le h(t_1) + \cdots + h(t_N)$, and in particular $h(t + ns) \le h(t) + nh(s)$ when n is a positive integer. Take $s = A_1/n$ and let $n \to \infty$. We obtain $h(t + A_1) \le h(t) + A_1 h'(0)$. Similarly, for $s = -A_2/n$, $h(t - A_2) \le h(t) - A_2 h'(1)$. A combination of these results yields the desired general inequality. The reader will convince himself that it is always strict except in trivial cases.

NOTES   It is difficult to trace the origins of harmonic measure, for the method was used much earlier than the name. For instance, the Lindelöf theorems in Sec. 3-2 were proved by a reasoning that comes very close to the two-constant theorem. Carleman [12], Ostrowski [49], and F. and R. Nevanlinna [41] used the method independently of each other. The name *harmonisches Mass* was introduced by R. Nevanlinna in his well-known monograph on analytic functions [45,46].

Beurling's proof of Theorem 3-6 is in his thesis [6] which appeared in 1933 and opened a whole new era in geometric function theory. The methods rather than the specific theorems in the thesis have been extremely influential.

The work of Teichmüller began to appear in the late thirties. Many of his articles were published in *Deutsche Mathematik* and are now very difficult to find.

EXERCISES

1   Let α be an arc on the boundary of a convex region. Show that the harmonic measure of α at a point z is at most $1/\pi$ times the angle subtended by α at z.

2   Let z_0 be a point in a Jordan region Ω and suppose (for simplicity) that the circle $|z - z_0| = R$ intersects the boundary at a finite number of points. Let f be analytic in Ω with $|f(z)| \le M$ everywhere and $|f(z)| \le m$ on the part of the boundary inside the circle. Show that

$|f(0)| \leq m^{\theta/2\pi} M^{1-\theta/2\pi}$, where θ is the total angular measure of the arcs on $|z - z_0| = R$ outside of Ω.

3   Let h_1 and h_2 be continuous functions in an interval $[a,b]$ such that $0 \leq h_1(x) \leq h_2(x)$. Let Ω be defined explicitly by $a < x < b$, $-h_1(x) < y < h_2(x)$. Show that the harmonic measure at a point on the real axis of the part of the boundary in the upper half plane is at most $\frac{1}{2}$.

4   In an isosceles triangle let $\omega(z)$ denote the harmonic measure of the base with respect to the triangle. If z moves on a line parallel to the base, show that $\omega(z)$ increases monotonically from a lateral side to the line of symmetry.

5   Let Ω be a region containing the origin and let α be part of its boundary. If n is a positive integer and S denotes rotation about the origin by the angle $2\pi/n$, assume that the component of 0 in $\Omega \cap S\Omega \cap \cdots \cap S^{n-1}\Omega$ has a boundary contained in $\alpha \cup S\alpha \cup \cdots \cup S^{n-1}\alpha$. Show that $\omega(0,\Omega,\alpha) \geq 1/n$.

4

EXTREMAL LENGTH

4-1 DEFINITION OF EXTREMAL LENGTH

In this chapter we discuss a geometric method that has had a profound
influence on the theory of conformal mapping, as well as on the more
general theory of quasiconformal mapping. It has its origin in an older
method, known as the *length-area principle*, which had been used occa-
sionally by several mathematicians and systematically above all by
Grötzsch. Briefly, the length-area principle uses euclidean length and
area, while the method of extremal length derives its higher degree of
flexibility and usefulness from more general ways of measuring.

Let Ω be a region in the plane, and Γ a set whose elements γ are
rectifiable arcs in Ω or, more generally, finite unions of such arcs (tech-
nically, each γ is a rectifiable one-chain). If Ω is mapped conformally on
Ω', the set Γ is transformed into a set Γ'. Our objective is to define a
number $\lambda_\Omega(\Gamma)$ with the invariance property $\lambda_\Omega(\Gamma) = \lambda_{\Omega'}(\Gamma')$.

It is natural to focus the attention on the lengths of the arcs γ.
However, length is not conformally invariant. For this reason we con-
sider the whole family of Riemannian metrics $ds = \rho|dz|$ which are con-

formal to the euclidean metric. This family is attached to the region Ω in a conformally invariant way. Indeed, under a conformal mapping $z \to z'$ the metric $\rho|dz|$ in Ω is transformed into a metric $\rho'|dz'|$ in Ω' with $\rho' = \rho|dz/dz'|$.

For technical reasons it is necessary to introduce some regularity requirements. This can be done in many ways, but we choose to require that the functions ρ be Borel measurable. In these circumstances every rectifiable arc γ has a well-defined ρ length

$$L(\gamma,\rho) = \int_\gamma \rho|dz|,$$

which may be infinite, and the open set Ω has a ρ area

$$A(\Omega,\rho) = \iint_\Omega \rho^2 \, dx \, dy.$$

If we perform a conformal mapping and replace ρ by ρ', as explained above, it is clear that $L(\Gamma,\rho) = L(\Gamma',\rho')$ and $A(\Omega,\rho) = A(\Omega',\rho')$. In order to define an invariant which depends on the whole set Γ, we introduce the minimum length

$$L(\Gamma,\rho) = \inf_{\gamma \in \Gamma} L(\gamma,\rho).$$

To obtain a quantity that does not change when ρ is multiplied by a constant we form the homogeneous expression $L(\Gamma,\rho)^2/A(\Omega,\rho)$. The set of all these ratios is conformally invariant, and so is their least upper bound. We are led to adopt the following definition.

Definition 4-1 The extremal length of Γ in Ω is defined as

$$\lambda_\Omega(\Gamma) = \frac{\sup_\rho L(\Gamma,\rho)^2}{A(\Omega,\rho)},$$

where ρ is subject to the condition $0 < A(\Omega,\rho) < \infty$.

There are several alternative statements of the definition obtained by use of different normalizations. For instance, $\lambda_\Omega(\Gamma)$ is equal to sup $L(\Gamma,\rho)^2$ when ρ is subject to the condition $0 < A(\Omega,\rho) \le 1$. Similarly, let us say that ρ is *admissible* if $L(\Gamma,\rho) \ge 1$ and define the *modulus* of Γ with respect to Ω as inf $A(\Omega,\rho)$ for admissible ρ. Then $\lambda_\Omega(\Gamma)$ is the reciprocal of the modulus, and it is a matter of taste whether one prefers to use the modulus or the extremal length. The modulus is denoted by $M_\Omega(\Gamma)$.

Another convenient normalization is expressed by the condition $L(\Gamma,\rho) = A(\Omega,\rho)$. Because of the different degrees of homogeneity it is always fulfilled for a suitable constant multiple of a given ρ, except when $L(\Gamma,\rho) = 0$ or ∞. With this normalization we have

$$\lambda_\Omega(\Gamma) = \sup L(\Gamma,\rho) = \sup A(\Omega,\rho). \tag{4-1}$$

Observe that $\lambda_\Omega(\Gamma) = 0$ if and only if $A(\Omega,\rho) < \infty$ implies $L(\Gamma,\rho) = 0$. In this case the normalization $L(\Gamma,\rho) = A(\Gamma,\rho)$ is possible only when both are zero.

The conformal invariance is an immediate consequence of the definition. We wish to point out, in addition, that in a sense $\lambda_\Omega(\Gamma)$ depends only on Γ and not on Ω. To see this, suppose that $\Omega \subset \Omega'$. Given ρ on Ω we choose $\rho' = \rho$ on Ω and $\rho' = 0$ on $\Omega - \Omega'$. Then ρ' is Borel measurable and $L(\Gamma,\rho) = L(\Gamma,\rho')$, $A(\Omega,\rho) = A(\Omega',\rho')$. This proves that $\lambda_{\Omega'}(\Gamma) \geq \lambda_\Omega(\Gamma)$. For the opposite inequality we need only start from a ρ' on Ω' and let ρ be its restriction to Ω. We see that $\lambda_\Omega(\Gamma)$ is the same for all open sets Ω that contain the arcs $\gamma \in \Gamma$. Accordingly, we shall henceforth simplify the notation to $\lambda(\Gamma)$.

4-2   EXAMPLES

To illustrate we consider the more special notion of *extremal distance*. Let Ω be an open set and let E_1, E_2 be two sets in the closure of Ω. Take Γ to be the set of *connected arcs* in Ω which join E_1 and E_2; in other words, each $\gamma \in \Gamma$ shall have one end point in E_1 and one in E_2, and all other points shall be in Ω. The extremal length $\lambda(\Gamma)$ is called the extremal distance of E_1 and E_2 in Ω, and we denote it by $d_\Omega(E_1, E_2)$. It depends essentially on Ω because the set Γ depends on Ω.

A typical example is the quadrilateral. A quadrilateral is a Jordan region Q together with four points on its boundary. These points divide the boundary into two pairs of opposite sides α, α' and β, β'. The quadrilateral is oriented by choosing one of these pairs, for instance α,α', as the base pair; the choice can be indicated by the notation $Q(\alpha,\alpha')$. We are interested in determining the extremal distance $d_Q(\alpha,\alpha')$.

Since the extremal distance is invariant under conformal mappings, we can replace Q by a conformally equivalent rectangle R. We choose the mapping so that the α sides of R lie on $x = 0$, $x = a$ and the β sides on $y = 0$, $y = b$. The first observation is that $\rho = 1$ gives $L(\Gamma,1) = a$, $A(R,1) = ab$. Hence $d_Q(\alpha,\alpha') \geq a^2/ab = a/b$. Conversely, let ρ be arbitrary but normalized by $L(\Gamma,\rho) = a$, for example. Then

$$\int_0^a [\rho(z) - 1]\, dx \geq 0,$$

and hence

$$\iint_R (\rho - 1)\, dx\, dy \geq 0.$$

Together with

$$\iint\limits_{R} (\rho - 1)^2 \, dx \, dy \geq 0$$

this leads to

$$A(R,\rho) = \iint\limits_{R} \rho^2 \, dx \, dy \geq \iint\limits_{R} dx \, dy = ab,$$

so that $L(\Gamma,\rho)^2/A(R,\rho) \leq a/b$ for all ρ. Hence $d_Q(\alpha,\alpha') \leq a/b$, and we have shown that the extremal distance between α and α' is equal to the ratio of the sides in a rectangle conformally equivalent to Q. If we interchange α and β, we obtain $d_Q(\beta,\beta') = b/a$. Observe that the product of the two extremal distances is 1.

There are other extremal lengths that can be associated with the configuration formed by Ω, E_1, and E_2. For instance, we could let Γ^* consist of all γ^* in Ω which separate E_1 and E_2; in this case we do not require each γ^* to be connected but allow it to be made up of several arcs or closed curves. The corresponding conformal invariant $\lambda(\Gamma^*)$ is called the *conjugate extremal distance* of E_1, E_2 with respect to Ω, and we denote it by $d_\Omega^*(E_1,E_2)$.

For instance, in the case of the quadrilateral it is quite evident that $d_Q^*(\alpha_1,\alpha_2) = d_Q(\beta_1,\beta_2)$. Thus the conjugate extremal distance is the reciprocal of the extremal distance. We shall find that this is the case in all sufficiently regular cases.

It is well known that every doubly connected region is conformally equivalent to an annulus $R_1 < |z| < R_2$. Therefore, if C_1 and C_2 denote the two components of the boundary of an annulus Ω, the extremal distance $d_\Omega(C_1,C_2)$ is the same as for an annulus. The reader will have no difficulty proving that $d_\Omega(C_1,C_2) = (1/2\pi) \log (R_2/R_1)$. The conjugate extremal distance $d_\Omega^*(C_1,C_2)$ is the extremal length of the family of closed curves that separate the contours, and its value is $2\pi\colon \log (R_2/R_1)$.

4-3   THE COMPARISON PRINCIPLE

The importance of extremal length derives not only from conformal invariance, but also from the fact that it is comparatively easy to find upper and lower bounds. First, any specific choice of ρ gives a lower bound for $\lambda_\Omega(\Gamma)$, namely, $\lambda_\Omega(\Gamma) \geq L(\Gamma,\rho)^2/A(\Omega,\rho)$. This may seem trivial, but it is nevertheless very useful.

To illustrate the point, let us apply this remark to a quadrilateral Q without first mapping on a rectangle. Let δ denote the shortest distance between α and α' inside Q, that is to say, the minimum euclidean length of an arc in Q which joins α and α'. If A denotes the area of Q, we obtain at once $d_Q(\alpha,\alpha') \geq \delta^2/A$. Similarly, $d_Q^*(\alpha,\alpha') \geq \delta^{*2}/A$, where δ^* is the shortest distance between β and β'. Since d_Q and d_Q^* are reciprocals, we even have a double inequality

$$\frac{\delta^2}{A} \leq d_Q(\alpha,\alpha') \leq \frac{A}{\delta^{*2}}.$$

From this we can derive the geometric inequality $\delta\delta^* \leq A$, which has no explicit connection with either extremal length or conformal mapping.

Besides these immediate estimates there is also a simple comparison principle which we shall formulate as a theorem.

Theorem 4-1 If every $\gamma \in \Gamma$ contains a $\gamma' \in \Gamma'$, then $\lambda(\Gamma) \geq \lambda(\Gamma')$.

Briefly, the set of fewer and longer arcs has the larger extremal length. The proof is a triviality. Indeed, both extremal lengths can be evaluated with respect to the same Ω. For any ρ in Ω it is clear that $L(\Gamma,\rho) \geq L(\Gamma',\rho)$. These minimum lengths are compared with the same $A(\Omega,\rho)$, and the assertion follows.

Corollary The extremal distance $d_\Omega(E_1,E_2)$ decreases when Ω, E_1, and E_2 increase.

However, the comparison principle is more general, even when applied only to extremal distances. Figure 4-1 shows the typical application to quadrilaterals: $d_{\Omega_1}(\alpha_1,\alpha_1') \geq d_{\Omega_2}(\alpha_2,\alpha_2')$.

4-4 THE COMPOSITION LAWS

In addition to the comparison principle there are two composition laws which express a relationship between three extremal lengths.

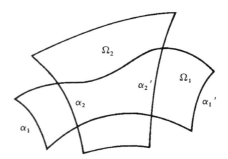

FIGURE 4-1

Theorem 4-2 Let Ω_1 and Ω_2 be disjoint open sets. Let Γ_1, Γ_2 consist of arcs in Ω_1, Ω_2, respectively, and let Γ be a third set of arcs.

(A) If every $\gamma \in \Gamma$ contains a $\gamma_1 \in \Gamma_1$ and a $\gamma_2 \in \Gamma_2$, then

$$\lambda(\Gamma) \geq \lambda(\Gamma_1) + \lambda(\Gamma_2). \tag{4-2}$$

(B) If every $\gamma_1 \in \Gamma_1$ and every $\gamma_2 \in \Gamma_2$ contains a $\gamma \in \Gamma$, then

$$1/\lambda(\Gamma) \geq 1/\lambda(\Gamma_1) + 1/\lambda(\Gamma_2). \tag{4-3}$$

PROOF   If $\lambda(\Gamma_1)$ or $\lambda(\Gamma_2)$ degenerates, being either 0 or ∞, the statements are trivial consequences of the comparison principle. To prove (A) in the nondegenerate case we choose ρ_1 in Ω_1 and ρ_2 in Ω_2, with the normalization $L(\Gamma_i,\rho_i) = A(\Omega_i,\rho_i)$, $i = 1$, 2. Consider an $\Omega \supset \Omega_1 \cup \Omega_2$ and take $\rho = \rho_1$ in Ω_1, $\rho = \rho_2$ in Ω_2, and $\rho = 0$ in $\Omega - \Omega_1 - \Omega_2$; this ρ is Borel measurable. We have trivially $L(\Gamma,\rho) \geq L(\Gamma_1,\rho_1) + L(\Gamma_2,\rho_2)$ and

$$A(\Omega,\rho) = A(\Omega_1,\rho_1) + A(\Omega_2,\rho_2) = L(\Gamma_1,\rho_1) + L(\Gamma_2,\rho_2).$$

Hence $\lambda(\Gamma) \geq L(\Gamma_1,\rho_1) + L(\Gamma_2,\rho_2)$, and (4-2) follows.

As for (B), let ρ be given in Ω and normalized by $L(\Gamma,\rho) = 1$. Then $L(\Gamma_1,\rho) \geq 1$, $L(\Gamma_2,\rho) \geq 1$, and this implies

$$A(\Omega,\rho) \geq A(\Omega_1,\rho) + A(\Omega_2,\rho) \geq \frac{1}{\lambda(\Gamma_1)} + \frac{1}{\lambda(\Gamma_2)}.$$

On the other hand, the greatest lower bound of $A(\Omega,\rho)$ is $1/\lambda(\Gamma)$, and (4-3) is proved.

The composition laws are best illustrated by the following simple examples:

In Fig. 4-2a, Ω is the interior of $\Omega' \cup \Omega'' \cup E$. Clearly, every arc in

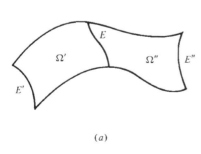
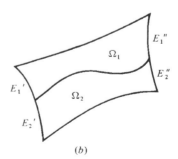

(a) (b)

FIGURE 4-2

Ω from E' to E'' contains an arc in Ω' from E' to E, and one in Ω'' from E to E''. Therefore, the first composition law implies

$$d_\Omega(E',E'') \geq d_{\Omega'}(E',E) + d_{\Omega''}(E,E''). \tag{4-4}$$

In Fig. 4-2b, Ω is the interior of $\bar{\Omega}_1 \cup \bar{\Omega}_2$. Every arc in Ω_1 from E'_1 to E''_1 and every arc in Ω_2 from E'_2 to E''_2 not only contains but actually is an arc in Ω from $E' = E'_1 \cup E'_2$ to $E'' = E''_1 \cup E''_2$. The second composition law yields

$$d_\Omega(E',E'')^{-1} \geq d_{\Omega_1}(E'_1,E''_1)^{-1} + d_{\Omega_2}(E'_2,E''_2)^{-1}. \tag{4-5}$$

In this case, because extremal distance and conjugate extremal distance are reciprocals, (4-3) and (4-4) happen to express the same fact, but this would not be so in more general circumstances.

4-5 AN INTEGRAL INEQUALITY

The composition laws can of course be applied to any finite number of regions. It is of interest to note that there is also an integrated counterpart of (4-4).

In Fig. 4-3, Ω represents a region between $x = a$ and $x = b$. We have denoted by $\theta(t)$ the length of the intercept with $x = t$. For small Δt it is evident that the extremal distance between the intercepts corresponding to t and $t + \Delta t$ is approximately $\Delta t/\theta(t)$. Therefore, the integrated version of (4-4) is

$$d_\Omega(E_1,E_2) \geq \int_a^b \frac{dx}{\theta(x)}. \tag{4-6}$$

To prove this inequality we need only choose $\rho(x,y) = 1/\theta(x)$. In this metric the length of any arc joining the vertical sides is at least equal

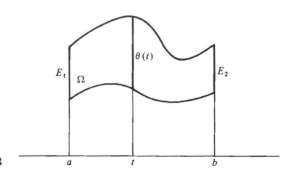

FIGURE 4-3

to $\int_a^b dx/\theta(x)$. But this integral is also the ρ area of Ω. Hence (4-6) is a direct consequence of the definition of extremal length. We have assumed for simplicity that each intercept consists of a single segment, and in that case $\theta(x)$ is lower semicontinuous so that ρ is certainly Borel measurable. If there are several segments, it is possible to choose $\theta(x)$ as the minimum length of the segments that join the lower and upper boundary curves.

The integral in (4-6) was first introduced by Ahlfors in a slightly different context. We shall return to this topic in Sec. 4-13.

4-6 PRIME ENDS

An important application of extremal length is to the boundary correspondence between two simply connected regions which are mapped conformally on each other. The classic theory, due to Carathéodory, is based on the concept of prime ends. We shall show in this section that the use of extremal length makes it possible to define prime ends in a conformally invariant manner. In terms of this definition the main theorem of Carathéodory will be a triviality. Naturally, it will be necessary to prove that the invariant definition is equivalent to the original one. This fact will emerge as a simple consequence of the comparison principle.

Let Ω be a simply connected region in the plane. A *crosscut* of Ω is a Jordan arc γ in Ω which in both directions tends to a boundary point. It is well known that $\Omega - \gamma$ consists of two simply connected components, and that γ constitutes the relative boundary of each component. We shall need the slightly more general notion of a *cluster of crosscuts*. Such a cluster is a finite union of crosscuts that form a connected point set. A cluster γ is said to separate two points in Ω if they are in different components of $\Omega - \gamma$.

Choose a fixed $z_0 \in \Omega$ and consider sequences $a = \{a_n\}$ of points in Ω. With the sequence a we associate the family Γ_a of all clusters of crosscuts of Ω which separate z_0 from almost all a_n, i.e., from all but a finite number of the a_n.

> **Definition 4-2** The sequence a is said to be fundamental if $\lambda(\Gamma_a) = 0$.

Recall that $\lambda(\Gamma_a) = 0$ if and only if $\inf_{\gamma \in \Gamma_a} L(\gamma, \rho) = 0$ for all ρ with $A(\Omega, \rho) < \infty$. In particular this condition is fulfilled for the spherical metric which we denote by ρ_0. We show at once that the definition is independent of the choice of z_0. Let z_0' be another choice and denote the corresponding family of clusters by Γ'_a. We connect z_0 to z'_0 by an arc c in Ω and let d be the shortest spherical distance from c to the boundary of Ω. Assume that $\lambda(\Gamma_a) = 0$. If $A(\Omega,\rho) < \infty$, it is also true that $A(\Omega,\rho + \rho_0) < \infty$. Therefore there exists a $\gamma \in \Gamma_a$ with arbitrarily small $L(\gamma,\rho + \rho_0)$. But then $L(\gamma,\rho)$ and $L(\gamma,\rho_0)$ are also arbitrarily small, and as soon as $L(\gamma,\rho_0) < d$, it is clear that $\gamma \in \Gamma'_a$. We have shown that $L(\Gamma'_a,\rho) = 0$, and hence that $\lambda(\Gamma'_a) = 0$.

We shall need an equivalence relation between fundamental sequences. For this purpose we consider the union $a \cup b$ of two sequences and show that the relation defined by $\lambda(\Gamma_{a \cup b}) = 0$ is transitive. In other words, if $a \cup b$ and $b \cup c$ are fundamental, we have to prove that $a \cup c$ is fundamental.

Suppose that $\gamma \in \Gamma_{a \cup b}$ and $\gamma' \in \Gamma_{b \cup c}$. If γ and γ' intersect, it is clear that the union $\gamma \cup \gamma'$ is again a cluster of crosscuts, and that it belongs to $\Gamma_{a \cup c}$. It is for the validity of this conclusion that we are forced to use clusters rather than individual crosscuts.

We assume now that γ and γ' do not intersect. In this situation γ lies in a component of $\Omega - \gamma'$, and γ' lies in a component of $\Omega - \gamma$. We denote by Ω_0 and Ω'_0 the components of $\Omega - \gamma$ and $\Omega - \gamma'$ which contain z_0. The following alternatives need to be considered.

1) γ' is not in Ω_0. With a finite number of exceptions a_n and c_n are not in Ω_0. Hence $\gamma \in \Gamma_{a \cup c}$.

2) γ' is in Ω_0. Choose b_n so that it is separated from z_0 by γ and γ'. Let Ω_1 and Ω'_1 be the components of $\Omega - \gamma$ and $\Omega - \gamma'$ which contain b_n. $\Omega_1 \subset \Omega'_1$, for otherwise Ω_1 would contain a relative boundary point of Ω'_1, hence a point on $\gamma' \subset \Omega_0$, contrary to the fact that Ω_0 and Ω_1 are disjoint. The situation is reduced to the previous case with the roles of γ and γ' reversed. Hence $\gamma' \in \Gamma_{a \cup c}$.

We have seen that in all circumstances $\gamma \cup \gamma'$, γ, or γ' belongs to $\Gamma_{a \cup c}$. It follows that $\lambda(\Gamma_{a \cup c}) = 0$, and the transitivity is established. The following definition is now meaningful.

Definition 4-3 The equivalence classes with respect to the relation $\lambda(\Gamma_{a \cup b}) = 0$ are called prime ends of Ω.

Let f be a conformal mapping from Ω to another region Ω'. Because of the conformal invariance of extremal length, it is immediately clear that the sequences $\{a_n\}$ and $\{f(a_n)\}$ are simultaneously fundamental. In other words, f determines a one-to-one correspondence between the prime ends of Ω and Ω'. This fact becomes highly significant as soon as we have a more direct way of recognizing fundamental sequences. Such a way is presented in the next theorem which states, in effect, that our definition coincides with Carathéodory's original definition.

Theorem 4-3 The sequence $\{a_n\}$ is fundamental if and only if a point $z_0 \in \Omega$ can be separated from almost all a_n by crosscuts of arbitrarily small spherical diameter.

As far as the necessity is concerned we have already remarked that z_0 is separated from almost all a_n by a cluster of crosscuts with arbitrarily small length, hence small diameter (all lengths and distances are with respect to the spherical metric). We wish to show that the separation can be accomplished by a single short crosscut. We need a lemma.

Lemma 4-1 Let Ω be a simply connected region and σ the union of a discrete set of mutually disjoint crosscuts γ. If two points $p, q \in \Omega$ are separated by σ, then they are also separated by one of the crosscuts γ.

PROOF Let Ω_0 be the component of $\Omega - \sigma$ that contains p. We join p to q by an arc c in Ω. Let p_0 be the last point of c on the boundary of Ω_0. It lies on one of the crosscuts, say γ_0. We claim that γ_0 separates p and q.

The discreteness means that every point on one of the crosscuts has a neighborhood which does not meet the others. Let Ω' be the component of $\Omega - \gamma_0$ that contains Ω_0. Because of the discreteness γ_0 is not in the relative boundary of $\Omega_0 \cup \gamma_0$ with respect to $\Omega' \cup \gamma_0$. Suppose that q were in Ω'. The set $\Omega_0 \cup \gamma_0$ is connected, p' lies in this set, and the subarc $p'q$ of c belongs to $\Omega' \cup \gamma_0$ without meeting the relative boundary of $\Omega_0 \cup \gamma_0$. It would follow that $q \in \Omega_0 \cup \gamma_0$, which is contrary to the assumption that σ separates p and q. We have shown that γ_0 is a separating crosscut.

PROOF OF THE THEOREM Let d be the distance from z_0 to the boundary of Ω and choose $\delta < d$. If $\{a_n\}$ is fundamental, there exists a cluster γ with diameter less than δ which separates z_0 from almost all a_n. Let C be a circle with radius δ whose center is an end point of γ. Then $\sigma = \Omega \cap C$ is a discrete set of disjoint crosscuts. According to the lemma one of these crosscuts separates z_0 from γ, hence from almost all a_n. Its diameter is at most 2δ and thus arbitrarily small.

For the converse, suppose that $\gamma_0 \in \Gamma_a$ is a crosscut of diameter $\delta < d$. With one end point as center we draw two circles C_1 and C_2 with radii δ and d. A simple closed curve which separates C_1 and C_2 will also separate z_0 from γ_0 in Ω. By the lemma it contains a crosscut which separates z_0 from γ_0 and thus from almost all a_n. This crosscut belongs to Γ_a, and the comparison principle permits us to conclude that $\lambda(\Gamma_a) \leq$

$d^*(C_1,C_2)$, the conjugate extremal distance between C_1 and C_2. If d and δ were euclidean radii, $d^*(C_1,C_2)$ would be equal to $2\pi : \log(d/\delta)$. The precise expression in terms of spherical radii hardly matters as long as it is clear that $d^*(C_1,C_2)$ tends to zero with δ. We conclude that $\lambda(\Gamma_a) = 0$ if δ can be chosen arbitrarily small.

If Ω is the unit disk, it follows at once from the theorem that a sequence is fundamental if and only if it converges to a point on the unit circle. In other words, the prime ends can be identified with the boundary points. It is not much harder to show, by very elementary topology, that the same is true for any Jordan region. Therefore, a conformal mapping of one Jordan region onto another can be extended to a homeomorphism between the closed regions. This is a basic theorem in conformal mapping.

For an arbitrary simply connected region Ω, other than the whole plane, a conformal mapping on the unit disk will define a one-to-one correspondence between the prime ends and the points on the unit circle. In particular, the prime ends have a natural cyclic order, and it is possible to speak of the cross ratio of four prime ends.

In order to further illustrate the use of prime ends, we consider the case of a Jordan region Ω with an incision c in the form of a Jordan arc from an interior point to a boundary point. The points of c are on the boundary of $\Omega - c$, and it is clear that in all reasonable applications the interior points of c should be counted twice. This becomes quite rigorous when we observe that there are indeed two distinct prime ends associated with each interior point of c. Boundary points with higher multiplicity have a similar interpretation.

In the general case the *cluster set* of a prime end consists of all the limits of convergent fundamental sequences in the equivalence class defined by the prime ends. It is easy to show that a cluster set is closed and

 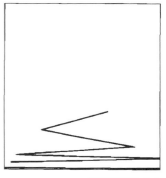

FIGURE 4-4

connected. Distinct prime ends may have the same cluster sets. Figure 4-4 shows examples of cluster sets that are not points; in the second example the cluster set belongs to two distinct prime ends.

4-7 EXTREMAL METRICS

We shall say that the metric ρ_0 is extremal for the family Γ in Ω if $L(\Gamma,\rho_0)^2/A(\Omega,\rho_0)$ is equal to its maximum $\lambda_\Omega(\Gamma)$. To compute an extremal length involves making a good guess what the extremal metric should be and then proving that the metric is in fact extremal.

The problem can be reversed: If ρ_0 is given, for what families of curves is it extremal? In unpublished work Beurling has given the following elegant and useful criterion.

Theorem 4-4 The metric ρ_0 is extremal for Γ if Γ contains a subfamily Γ_0 with the following properties:

$$i) \qquad \int_\gamma \rho_0|dz| = L(\Gamma,\rho_0) \qquad \text{for all} \qquad \gamma \in \Gamma_0; \qquad (4\text{-}7)$$

$ii)$ for real-valued h in Ω the conditions

$$\int_\gamma h|dz| \geq 0 \qquad (4\text{-}8)$$

for all $\gamma \in \Gamma_0$ imply

$$\iint_\Omega h\rho_0 \, dx \, dy \geq 0. \qquad (4\text{-}9)$$

PROOF The proof is almost trivial. Let ρ be normalized by

$$L(\Gamma,\rho) = L(\Gamma,\rho_0).$$

Then
$$\int_\gamma \rho|dz| \geq \int_\gamma \rho_0|dz|$$

for all $\gamma \in \Gamma_0$, so that (4-8) is fulfilled with $h = \rho - \rho_0$. It follows by (4-9) that

$$\iint_\Omega (\rho\rho_0 - \rho_0{}^2) \, dx \, dy \geq 0,$$

and an obvious application of the Schwarz inequality gives

$$\iint_\Omega \rho_0{}^2 \, dx \, dy \leq \iint_\Omega \rho^2 \, dx \, dy,$$

proving that ρ_0 is extremal. [The proof indicates what regularity to impose on h and how to interpret (4-8) and (4-9).]

EXAMPLE 4-1 For the extremal distance between the vertical sides of a rectangle $R = \{a < x < b, \; c < y < d\}$ we take $\rho_0 = 1$ and let Γ_0 be formed by the lines $y = $ constant. It is evident by integration that

$$\int_a^b h(x,y) \, dx \geq 0$$

implies

$$\iint_R h(x,y) \, dx \, dy \geq 0.$$

Beurling's criterion is satisfied, and $\rho_0 = 1$ is extremal.

EXAMPLE 4-2 As a less obvious example we consider a triangle or, conformally speaking, a Jordan region with three distinguished boundary points. Since all such configurations are conformally equivalent, all extremal length problems associated with this situation will lead to specific numbers rather than conformal invariants. As a particular instance we shall determine $\lambda(\Gamma)$, where Γ consists of all arcs in the triangle that touch all three sides.

We begin by mapping conformally on an equilateral triangle with side 1, and we shall show that $\rho_0 = 1$, the euclidean metric, is extremal. Reflection in a side (Fig. 4-5a) shows that the minimum length of $\gamma \in \Gamma$ is that of the altitude. The shortest arcs are broken lines which make up three subfamilies, one of which is shown in Fig. 4-5b. We take Γ_0 to be the family of all these broken lines. Let γ_x denote the arc in Γ_0 which begins at $(x,0)$, $0 \leq x \leq \frac{1}{2}$. If h satisfies condition (4-8), integration gives

$$\int_0^{\frac{1}{2}} dx \int_{\gamma_x} h \, |dz| = \iint h \, dx \, dy \geq 0,$$

where the double integral is taken once over the shaded area and twice over the doubly shaded area. There are three such integrals, and when the results are added up we obtain

$$3 \iint h \, dx \, dy \geq 0,$$

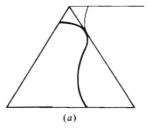
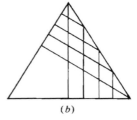

(a) $\qquad\qquad\qquad\qquad\qquad$ (b)

FIGURE 4-5

where the integral is now over the whole triangle. We conclude by the theorem that ρ_0 is extremal, and computation yields $\lambda(\Gamma) = \sqrt{3}$.

4-8 A CASE OF SPHERICAL EXTREMAL METRIC

In the examples above the extremal metric was euclidean, so that $\rho_0 = 1$ after a preliminary conformal mapping. We shall now discuss an example in which this is not the case.

In a Jordan region Ω let z_0 be an interior point and let Γ be the family of Jordan curves in Ω that touch the boundary and enclose z_0 (more accurately, $\gamma \in \Gamma$ becomes a Jordan curve when one boundary point is added). By conformal invariance we can take Ω to be the unit disk Δ, with $z_0 = 0$.

We shall first compare $\lambda(\Gamma)$ with $\lambda(\Gamma_1)$, where Γ_1 consists of the arcs in Δ which join diametrically opposite boundary points. The comparison is by means of the mapping $z = z_1{}^2$. Clearly, every $\gamma_1 \in \Gamma_1$ is mapped on a $\gamma \in \Gamma$, and every γ is the image of two γ_1. Given ρ we determine ρ_1 so that $\rho_1|dz_1| = \rho|dz|$; this means setting $\rho_1(z_1) = 2|z_1|\rho(z_1{}^2)$. With this choice it is immediate that $L(\Gamma,\rho) = L(\Gamma_1,\rho_1)$ and $2A(\Delta,\rho) = A(\Delta,\rho_1)$. It follows that $\lambda(\Gamma) \leq 2\lambda(\Gamma_1)$.

In the opposite direction, if ρ_1 is given, a single-valued ρ can be defined by $\rho(z) = \frac{1}{4}|z|^{-\frac{1}{2}}[\rho_1(z^{\frac{1}{2}}) + \rho_1(-z^{\frac{1}{2}})]$. It is readily seen that $L(\Gamma,\rho) \geq L(\Gamma_1,\rho_1)$ and $2A(\Delta,\rho) \leq A(\Delta,\rho_1)$. This proves $\lambda(\Gamma) \geq 2\lambda(\Gamma_1)$, so that in fact $\lambda(\Gamma) = 2\lambda(\Gamma_1)$.

In order to determine the extremal metric for Γ_1 we map Δ by stereographic projection on the upper half of the Riemann sphere with radius 1 and center at the origin. The points on the unit circle are their own images, and the mapping is conformal. Although the hemisphere is not a plane region, it is clear that the method of extremal length is applicable with insignificant changes, and that Theorem 4-4 remains valid if the integrals are taken with respect to the spherical metric. We shall show that the spherical metric is extremal. This is the same as saying that $\rho_0 = 2(1 + |z|^2)^{-1}$ is extremal for the original problem.

The shortest curves in the spherical metric are the great circles; we must show that they satisfy condition (ii) in Theorem 4-4. We shall use geographic coordinates θ for longitude and φ for latitude. Consider a great circle whose inclination is determined by the maximal value φ_0 of the latitude. We keep φ_0 fixed and rotate about the vertical axis; the semicircles sweep out the zone $0 \leq \varphi \leq \varphi_0$. We leave it to the reader to verify that

$$ds\, d\theta = (\cos^2 \varphi - \cos^2 \varphi_0)^{-\frac{1}{2}}\, d\omega,$$

where s denotes arc length along the great circle and $d\omega$ is the spherical area element. If h satisfies

$$\int_{\gamma_0} h \, ds \geq 0 \tag{4-10}$$

for all great circles γ_0, it follows that

$$\iint_{\varphi < \varphi_0} (\cos^2 \varphi - \cos^2 \varphi_0)^{-\frac{1}{2}} h \, d\omega \geq 0. \tag{4-11}$$

Now multiply with $\sin \varphi_0$ and integrate from $\varphi_0 = 0$ to $\varphi_0 = \pi/2$. Because

$$\int_0^{\pi/2} (\cos^2 \varphi - \cos^2 \varphi_0)^{-\frac{1}{2}} \sin \varphi_0 \, d\varphi_0 = \frac{\pi}{2},$$

it follows that

$$\iint h \, d\omega \geq 0$$

over the hemisphere. This proves our contention, and we find $\lambda(\Gamma_1) = \pi/2$, $\lambda(\Gamma) = \pi$.

It is of some interest that the same method solves a more general problem. Suppose that we know (4-10) only for those great circles whose maximal latitude φ_0 belong to an interval $[0, \varphi_1]$ with $\varphi_1 < \pi/2$. Then (4-11) is valid for these values of φ_0. This time we multiply by the factor $\cos \varphi_0 \sin \varphi_0 (\cos^2 \varphi_0 - \cos^2 \varphi_1)^{-\frac{1}{2}}$ and integrate with respect to φ_0 from 0 to φ_1. The substitution $t_0 = \cos \varphi_0$ yields

$$\int_{\varphi}^{\varphi_1} [(\cos^2 \varphi_0 - \cos^2 \varphi_1)(\cos^2 \varphi - \cos^2 \varphi_0)]^{-\frac{1}{2}} \cos \varphi_0 \sin \varphi_0 \, d\varphi_0$$

$$= \int_{t_1}^{t} [(t_0^2 - t_1^2)(t^2 - t_0^2)]^{-\frac{1}{2}} t_0 \, dt_0 = \frac{\pi}{2}$$

independently of φ and φ_1. We conclude that

$$\iint_{0 < \varphi < \varphi_1} h \, d\omega \geq 0,$$

and hence that the spherical metric is extremal in any zone $0 \leq \varphi \leq \varphi_1$.

It is thereby clear that we have determined the extremal length of the family of arcs that join opposite points on the outer boundary of an annulus. More important, by the same comparison as in the case of the punctured disk we can find the extremal length of the closed curves that touch the outer boundary and enclose the inner boundary of a doubly connected region. We leave it to the reader to compute the numerical value.

4-9 THE EXPLICIT FORMULA FOR EXTREMAL DISTANCE

We return to the extremal distance $d_\Omega(E_1,E_2)$ introduced in Sec. 4-2. Our aim is to find an explicit expression for this conformal invariant, at least when the configuration formed by Ω, E_1, and E_2 is fairly simple. There is no generality lost in assuming that E_1 and E_2 are contained in the boundary of Ω, for this will be the case if we replace Ω by a component of $\Omega - (E_1 \cup E_2)$. To avoid delicate discussions of the boundary behavior, we shall make the following assumptions: (1) Ω is a bounded region whose boundary consists of a finite number of Jordan curves; (2) E_1 and E_2 are disjoint, and each is a finite union of closed arcs or closed curves contained in the boundary of Ω. Under these conditions there exists a conformal mapping of Ω on a region with analytic boundary curves, and the mapping extends to a homeomorphism of the closed regions. Since our problem is invariant under conformal mapping, we may as well assume from the start that Ω has an analytic boundary.

We denote the full boundary of Ω by C, and we set $C_0 = C - (E_1 \cup E_2)$. Also, $E_1{}^0$ and $E_2{}^0$ will denote the relative interiors of E_1 and E_2 as subsets of C; they are obtained by removing the end points. There exists a unique function $u(z)$ in Ω with the following properties:

  i) u is bounded and harmonic in Ω,
  ii) u has a continuous extension to $\Omega \cup E_1{}^0 \cup E_2{}^0$, which is equal to 0 on $E_1{}^0$ and 1 on $E_2{}^0$,
  iii) the normal derivative $\partial u/\partial n$ exists and vanishes on C_0.

This function is the solution of a mixed Dirichlet-Neumann problem. The uniqueness follows from the maximum principle which also guarantees that $0 < u < 1$ in Ω. The existence will be taken for granted. The reflection principle implies that u has a harmonic extension across $C_0 \cup E_1{}^0 \cup E_2{}^0$.

Theorem 4-5   The extremal distance $d_\Omega(E_1,E_2)$ is the reciprocal of the Dirichlet integral

$$D(u) = \iint\limits_\Omega (u_x{}^2 + u_y{}^2) \, dx \, dy.$$

PROOF   If we rely on geometric intuition, the proof is almost trivial. Choose $\rho_0 = |\text{grad } u| = (u_x{}^2 + u_y{}^2)^{\frac{1}{2}}$, and let γ be an arc from E_1 to E_2. Since $|\text{grad } u|$ is the maximum of the directional derivative, we have at once

$$\int_\gamma \rho_0|dz| \geq \int_\gamma |du| \geq \int_\gamma du = 1,$$

and hence $L(\Gamma,\rho_0) \geq 1$ for the family Γ of joining arcs. On the other hand $A(\Omega,\rho_0) = D(u)$, and we conclude that $d_\Omega(E_1,E_2) \geq L(\Gamma,\rho_0)^2/A(\Omega,\rho_0) \geq 1/D(u)$.

It is the rigorous proof of the opposite inequality that causes some difficulty. If v denotes the conjugate harmonic function of u, it is fairly evident that the lines on which v is constant are the shortest arcs in the metric ρ_0. The trouble is that v need not be single-valued, and for this reason the level lines are defined only locally. There is one such line passing through each point z with grad $u(z) \neq 0$, but the level lines branch at the *critical points*, namely, the points where grad $u = 0$. We shall see that there are only a finite number of critical points, but even with this information it is not at all obvious what the global behavior of the level lines will be.

Postponing these difficulties, let us assume that Ω can be swept out by level lines $v = t$ which pass from E_1 to E_2. We must allow for a finite number of sudden changes of t, but apart from these jumps the total increase in t will be

$$\int_{E_1} dv = -\int_{E_1} \frac{\partial u}{\partial n}\,|dz|,$$

where n is the outer normal. By Green's theorem the integral of $\partial u/\partial n$ over the whole boundary is zero. Hence we obtain

$$-\int_{E_1} \frac{\partial u}{\partial n}\,|dz| = \int_{E_2} \frac{\partial u}{\partial n}\,|dz| = \int_C u\,\frac{\partial u}{\partial n}\,|dz| = D(u),$$

so that the increase in t is $D(u)$.

Along a level line $\rho_0 = \partial u/\partial s$, and the ρ length of the level line can be expressed as

$$\int_{v=t} \rho|dz| = \int_{v=t} \frac{\rho}{\rho_0}\,du.$$

If ρ is normalized by $L(\Gamma,\rho) = 1$ we thus have

$$\int_{v=t} \frac{\rho}{\rho_0}\,du \geq 1,$$

and integration with respect to t yields

$$\iint \frac{\rho}{\rho_0}\,du\,dv \geq D(u),$$

where the integral is over a union of rectangles. But $du\,dv = \rho_0{}^2\,dx\,dy$, and we thus have

$$\iint\limits_{\Omega} \rho\rho_0 \, dx \, dy \geq D(u).$$

The Schwarz inequality gives

$$D(u) \leq \iint\limits_{\Omega} \rho^2 \, dx \, dy = A(\Omega,\rho),$$

and if the reasoning can be trusted we have shown that $d_\Omega(E_1,E_2) \geq 1/D(u)$.

Before proceeding to a more rigorous proof we illustrate the method by some examples. Consider first a triply connected region Ω and let E_2 consist of the outer contour while E_1 is composed of the two inner contours (Fig. 4-6).

In this case u is the ordinary harmonic measure, $u = 0$ on E_1, and $u = 1$ on E_2. For small positive ϵ the level curve $u = \epsilon$ will be a pair of simple closed curves near E_1, while $u = 1 - \epsilon$ will be a single closed curve near E_2. There is obviously a critical value u_0 such that the level curve $u = u_0$ is a figure-eight-shaped curve with a point of self-intersection. This point is a critical point of u in the sense that $u_x = u_y = 0$.

The figure-eight-shaped curve divides the region into three parts, each doubly connected, with $u < u_0$ in two of the regions and $u > u_0$ in the third. As indicated in Fig. 4-6 we obtain a model of the triply connected region in the form of a rectangle cut up into three subrectangles and with identifications which make it resemble a pair of trousers. The total area is $D(u)$.

We now consider a slightly more complicated case. The region will again be triply connected and E_1 will consist of the inner contours, but

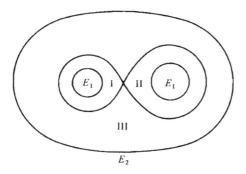
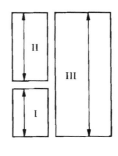

FIGURE 4-6

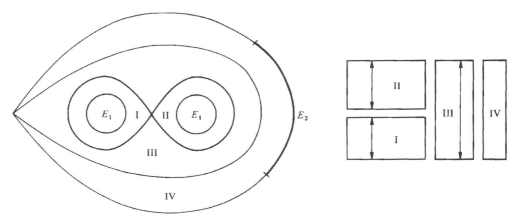

FIGURE 4-7

this time E_2 will be only a subarc of the outer contour (Fig. 4-7). The family of level curves is similar to the one in the preceding example, but there will be an additional critical point on the boundary. As before we draw the level curves through the critical points and observe how they divide the region. There are three annular and one simply connected subregion. The latter is mapped on a rectangle whose horizontal sides are not identified. The model resembles an unbuttoned pair of trousers.

We are now ready for a valid proof in the general case. First, the critical points are zeros of the analytic function $u_x - iu_y$. We have already remarked that u has a harmonic extension across $C_0 \cup E_1 \cup E_2$. Therefore $u_x - iu_y$ has an analytic extension from which it follows that its zeros, the critical points, have no accumulation points except possibly the end points of E_1 and E_2. Let z_0 be such an end point, for instance, of E_1. We can choose a local conjugate v in Ω near z_0 such that, on the boundary, $u = 0$ on one side of z_0 and $v = 0$ on the other side of z_0. Then $(u + iv)^2$ is real on both sides of z_0, and by the reflection principle there exists a neighborhood V of z_0 and an analytic function φ in $V - \{z_0\}$ such that $\varphi = (u + iv)^2$ in $\Omega \cap V$. Moreover, the real part of φ is symmetric across the boundary. Since $\operatorname{Re} \varphi = u^2 - v^2 < 1$ in $\Omega \cap V$, the same is true throughout $V - \{z_0\}$. This proves that z_0 is neither a pole nor an essential singularity of φ. Consequently, $\varphi(z_0)$ exists as a limit, and from the fact that u does not change its sign in Ω we are able to conclude that φ has a simple zero at z_0. Therefore $\varphi'(z_0) = 2(u + iv)(u_x - iu_y) \neq 0$, and $u_x - iu_y$ must tend to ∞ as $z \to z_0$. Hence there are no critical points near z_0, and the number of critical points is finite.

This having been established, let u_i be the values of u at the critical points in increasing order. We examine the subregions of Ω characterized by $u_i < u(z) < u_{i+1}$, where, for example, $u_0 = 0$, $u_n = 1$. Actually, these subregions may not be connected, in which case we consider the components separately. We are going to show that each component is of the type considered in our examples.

For this purpose we need to compute the number of critical points. Let us assume that Ω has m contours, and that E_1 and E_2 consist of arcs with a total of h end points. The critical points are the zeros of $u_x - iu_y$, each zero having a certain multiplicity. Let there be n_1 critical points in the interior and n_2 on the boundary when counted with these multiplicities.

We use the generalized form of the argument principle in which zeros and poles on the boundary are counted with half multiplicity. This means that

$$\int_C d \arg (u_x - iu_y) = \left(n_1 + \frac{1}{2} n_2 - \frac{h}{4} \right) 2\pi,$$

where the term $-h/4$ is due to the fact that $u_x - iu_y$ must be regarded as having half a pole at each end point of the E_1 and E_2. Indeed, if z_0 is an end point, $u + iv \sim (z - z_0)^{\frac{1}{2}}$ and $u_x - iu_y \sim \frac{1}{2}(z - z_0)^{-\frac{1}{2}}$.

We allow ourselves to write $w = u + iv$ and $u_x - iu_y = dw/dz$ even though w is not single-valued. It becomes clear that

$$\int_C d \arg (u_x - iu_y) = \int_C d \arg dw - \int_C d \arg dz$$

in obvious notation. The first integral on the right is zero because $\arg dw$ is constant on each subarc of C. The second integral measures the turning of the tangent and is therefore equal to $(2 - m)2\pi$. When the results are combined, we obtain

$$2n_1 + n_2 = 2m - 4 + \frac{h}{2}.$$

In particular, if there are no critical points, we must have $2m + h/2 = 4$. This allows for only two possibilities: (1) $m = 1$, $h = 4$; (2) $m = 2$, $h = 0$. The region is either a quadrilateral or an annulus, and these are precisely the cases that were considered above.

We apply the result to a component of $u_i < u < u_{i+1}$. There are no critical points inside, nor on the part of the boundary of the component that lies on $E_1 \cup E_2$. There may be, and usually are, critical points on the boundary, but at these points the boundary forms an angle, and the preliminary transformation needed to straighten this angle will cancel out the multiplicity of the critical point. The net effect is that our formula is still valid with $n_1 = n_2 = 0$, and we conclude that all subregions are quadrilaterals or annuli. The components of $u_i < u < u_{i+1}$ can therefore be mapped conformally on rectangles of width $u_{i+1} - u_i$ and combined height $D(u)$. Together they fill out a rectangle with sides 1 and $D(u)$. After appropriate identifications we obtain a model of Ω with E_1 and E_2 as vertical sides. From this model it is immediately clear that the euclidean metric is extremal, and we conclude that $d_\Omega(E_1, E_2) = 1/D(u)$.

4-10 CONFIGURATIONS WITH A SINGLE MODULUS

The term *configuration* will be used to designate a region bounded by a finite number of smooth curves together with a finite set of interior and boundary points taken in a certain order. Two configurations are conformally equivalent if there exists a conformal mapping of one region on the other which maps the specified interior and boundary points on the corresponding points of the other region. The equivalence of configurations can be expressed through the equality of certain conformal invariants called *moduli*. These invariants can be chosen in various way, but their number is always the same.

The number of moduli can be determined by mapping on a canonical region, for instance, on a parallel slit region. If the configuration has m contours, n_1 interior points, and n_2 boundary points, it turns out that the number of moduli is

$$N = 3m - 6 + 2n_1 + n_2 + s, \tag{4-12}$$

where $s = 0$ except in five cases. Actually, s is the number of parameters in the family of conformal self-mappings. It is 3 for a disk, 2 for a disk with one boundary point, and 1 for an annulus and for a disk with one interior or two boundary points.

We shall not be concerned with the proof or interpretation of formula (4-12). Our aim is rather to undertake a somewhat deeper study of the configurations with a single modulus. With $s = 0$ this will happen when $3m + 2n_1 + n_2 = 7$. The possible cases are

$$
\begin{array}{llll}
(1) & m = 1 & n_1 = 2 & n_2 = 0 \\
(2) & m = 1 & n_1 = 1 & n_2 = 2 \\
(3) & m = 1 & n_1 = 0 & n_2 = 4 \\
(4) & m = 2 & n_1 = 0 & n_2 = 1.
\end{array}
$$

In addition there is the case of the annulus with $s = 1, m = 2, n_1 = n_2 = 0$.

Case 1   Simply connected region with two interior points   We already have a conformal invariant, namely, the Green's function $g(z_1, z_2)$. Every other invariant must be a function of this one. It is nevertheless instructive to form other invariants and to compare them with the Green's function.

Case 2   Simply connected region with one interior and two boundary points   A conformal invariant is the harmonic measure at the interior point of the arc between the boundary points. The same remark applies as above.

Case 3   The quadrilateral   We have already considered its modulus, which is the extremal distance between a pair of opposite sides. Another invariant can be obtained by mapping on a disk or half plane and forming the cross ratio of the four points.

Case 4   Annulus with one boundary point   The boundary point is of no interest since we can move it to any position by a rotation. Therefore this is just the case of the annulus, and the modulus is the extremal distance between the contours.

4-11   EXTREMAL ANNULI

We take a closer look at Case 1 of the preceding section. Assume that the region is a disk and that z_1, z_2 are on a diameter. As a modulus we introduce $\lambda^* = \lambda(\Gamma^*)$, where Γ^* is the family of closed curves that separate z_1 and z_2 from the circumference. Let s be the line segment joining z_1 to z_2 and denote by Γ_0^* the smaller family of closed curves that separate s from the circumference. Now we are dealing with an annulus, and we denote the conjugate extremal distance between s and the circle by $\lambda_0^* = \lambda(\Gamma_0^*)$. There is an extremal metric ρ_0 for the family Γ_0^*, which is obviously symmetric with respect to s.

Given $\gamma^* \in \Gamma^*$ we obtain a γ_0^* of equal ρ_0 length by reflecting part of γ^* across s. Although γ_0^* is not strictly contained in the annulus, it is clear that $L(\Gamma^*,\rho_0) = L(\Gamma_0^*,\rho_0)$, and we conclude that $\lambda^* = \lambda_0^*$. Note the crucial role of the symmetry.

We can look at this result a little differently. Let c be any continuum that contains z_1 and z_2. Let d_c be the extremal distance from c to the circle. Then the maximum of d_c is clearly a conformal invariant. On passing to conjugate extremal distances we have

$$d_c^* \geq \lambda^* = \lambda_0^* = d_s^*,$$

and hence $d_c \leq d_s$. In other words, of all continua containing z_1 and z_2 the line segment is "farthest away" from the circle.

In view of conformal invariance the following is an equivalent statement:

Theorem 4-6 (Grötzsch) Of all the continua that join the point $R > 1$ to ∞ the segment $[R, +\infty]$ of the real axis has the greatest extremal distance from the unit circle.

The doubly connected region whose complement consists of the closed unit disk and the segment $[R, +\infty]$ is known as the Grötzsch annulus. We shall denote its modulus, i.e., the extremal distance between the components of the complement, by $M(R)$. This means that the Grötzsch annulus is conformally equivalent to a circular annulus whose radii have the ratio $e^{2\pi M(r)}$.

Another extremal problem of similar nature was solved by Teichmüller [65].

Theorem 4-7 (Teichmüller) Of all doubly connected regions that separate the pair $\{0, -1\}$ from a pair $\{w_0, \infty\}$ with $|w_0| = R$ the one with the greatest modulus is the complement of the segments $[-1, 0]$ and $[R, +\infty]$.

The proof makes use of Koebe's one-quarter theorem (Sec. 2-3) and Koebe's distortion theorem (to be proved in Chap. 5). Suppose that f is univalent in the unit disk and normalized by $f(0) = 0$, $f'(0) = 1$. The one-quarter theorem states that $f(z) \neq w_0$ for $|z| < 1$ implies $|w_0| \geq \frac{1}{4}$. The distortion theorem asserts in part that

$$|f(z)| \leq \frac{|z|}{(1 - |z|)^2}.$$

In both cases there will be equality for the Koebe function

$$f_1(z) = \frac{z}{(1 + z)^2}.$$

In fact, $f_1(z)$ does not assume the value $\frac{1}{4}$ and $|f_1(z)| = |z|/(1 - |z|)^2$ for negative z.

Let Ω be the doubly connected region in the theorem, and E_1 the bounded and E_2 the unbounded component of the complement. The set $\Omega \cup E_1$ is a simply connected region, and not the whole plane. By Riemann's mapping theorem there exists a univalent function F in the unit disk $\Delta = \{|z| < 1\}$ such that $F(\Delta) = \Omega \cup E_1$ and $F(0) = 0$. Because

$F(z) \neq w_0$, we may conclude by the one-quarter theorem that

$$R = |w_0| \geq \tfrac{1}{4}|F'(0)|.$$

Let z_0 be the inverse image of -1. The distortion theorem yields

$$1 = |F(z_0)| \leq \frac{|z_0|\,|F'(0)|}{(1 - |z_0|)^2} \leq \frac{4R|z_0|}{(1 - |z_0|)^2}. \tag{4-13}$$

Suppose now that Ω is the Teichmüller annulus with $E_1 = [-1, 0]$ and that $E_2 = [R, +\infty]$. The mapping function is $F_1(z) = 4Rf_1(z)$, where f_1 is the Koebe mapping, and $z_1 = F_1^{-1}(-1) = f_1^{-1}(-\tfrac{1}{4}R)$, which is negative. Hence there is equality in both places in (4-13) when F is replaced by F_1 and z_0 by z_1. We conclude that $|z_0| \geq |z_1|$, for $t/(1 - t)^2$ is an increasing function.

We note further that the modulus of the original Ω is the extremal distance between $F^{-1}(E_1)$ and the unit circle. Given that 0 and z_0 belong to $F^{-1}(E_1)$ we already know that this extremal distance is greatest when $F^{-1}(E_1)$ is the line segment between 0 and z_0. The line segment is shorter and the extremal distance greater when z_0 is replaced by z_1. Hence the modulus is indeed a maximum for the Teichmüller annulus.

For the convenience of the reader Fig. 4-8 shows the Grötzsch and Teichmüller annuli. It is clear that the Grötzsch domain together with its

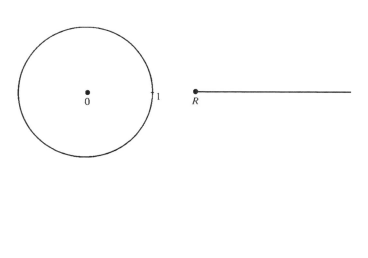

FIGURE 4-8

reflection in the circle is a Teichmüller domain, except for normalization. If the modulus of the Teichmüller annulus is denoted by $\Lambda(R)$, it follows readily that M and Λ are connected by the relation

$$\Lambda(R^2 - 1) = 2M(R). \qquad (4\text{-}14)$$

The nature of the functions $M(R)$ and $\Lambda(R)$ will be discussed in the next section. For the moment we remark only that the value $\Lambda(1)$ can be found at once. If $R = 1$, it is in fact obvious that the part of the Teichmüller annulus in the upper half plane is conformally equivalent to a square. Hence the extremal distance is 1 with respect to the half plane and $\frac{1}{2}$ with respect to the full plane. Thus $\Lambda(1) = \frac{1}{2}$, a special case of the more general relation $\Lambda(R)\Lambda(1/R) = \frac{1}{4}$, the proof of which is left as an exercise.

Corollary Every doubly connected region with modulus $>\frac{1}{2}$ contains a circle which separates the components of the complement.

For the proof we may assume that the bounded component E_1 has diameter 1 and that the points 0 and -1 belong to E_1. If the assertion were not true, E_2 would meet the circle $|z| = 1$, and by the theorem the modulus would be $\leq \Lambda(1) = \frac{1}{2}$.

4-12 THE FUNCTION $\Lambda(R)$

The theory of elliptic functions makes it possible to find an explicit expression for $\Lambda(R)$, or rather for its inverse function. Such computations are not as popular as they used to be, and for the convenience of the reader we shall reproduce some details.

Recall that the Weierstrass $\wp$ function with period basis ω_1, ω_2 is defined by

$$\wp(z) = \frac{1}{z^2} + \sum_{\omega \neq 0} \left(\frac{1}{(z - \omega)^2} - \frac{1}{\omega^2} \right), \qquad (4\text{-}15)$$

where ω ranges over all periods except 0. It satisfies the differential equation

$$\wp'(z)^2 = 4[\wp(z) - e_1][\wp(z) - e_2][\wp(z) - e_3], \qquad (4\text{-}16)$$

with $e_1 = \wp(\omega_1/2)$, $e_2 = \wp(\omega_2/2)$, $e_3 = \wp[(\omega_1 + \omega_2)/2]$. This relation is proved by comparing the zeros and poles on both sides of (4-16). In particular, the values e_1, e_2, e_3 are assumed with multiplicity two. It follows that they are distinct, for if two were equal p would assume the same value four times.

We shall choose $\omega_1 = 1$ and $\omega_2 = 2i\Lambda$, with $\Lambda = \Lambda(R)$, the modulus of the Teichmüller annulus as described in the preceding section. It is quite obvious from (4-15) that the $\wp$ function is real on the real and imaginary axes and on their parallels translated by half periods. Thus e_1, e_2, e_3 are real, and the $\wp$ function maps the perimeter of the rectangle with vertices 0, $\omega_1/2$, $(\omega_1 + \omega_2)/2$, $\omega_2/2$ on the real axis. It is easy to see that the rectangle is in one-to-one correspondence with either the upper or the lower half plane. Examination of the behavior near $z = 0$ shows that it is the lower half plane. The points ∞, e_1, e_3, e_2 must follow each other in positive order with respect to the lower half plane so that $e_2 < e_3 < e_1$. The segments $[e_1, +\infty]$ and $[e_2, e_3]$ correspond to the horizontal sides of the rectangle and thus have the extremal distance 2Λ with respect to the half plane and Λ with respect to the full plane. Hence ∞, e_1, e_3, e_2 correspond by linear transformation to $\infty, R, 0, -1$, so that

$$R = \frac{e_1 - e_3}{e_3 - e_2}. \tag{4-17}$$

Our task is to express R in terms of Λ. For this purpose it is sufficient to construct an elliptic function F with the same periods, zeros, and poles as $[\wp(z) - e_1]/[\wp(z) - e_2]$. Indeed, it is then clear that

$$R = -\frac{F[(\omega_1 + \omega_2)/2]}{F(0)}.$$

It is convenient to use the notation $q = e^{-2\pi\Lambda}$. We shall verify that the function

$$F(z) = \prod_{n=-\infty}^{\infty} \frac{(1 + q^{2n}e^{-2\pi iz})^2}{(1 - q^{2n-1}e^{-2\pi iz})(1 - q^{2n+1}e^{-2\pi iz})} \tag{4-18}$$

has the desired properties. First, the product converges at both ends because $q < 1$ and because each factor remains unchanged when n is replaced by $-n$ and z by $-z$. It has trivially the period 1, and replacing z by $z + 2i\Lambda$ amounts to replacing n by $n - 1$ so that $2i\Lambda$ is also a period. Finally, the zeros and poles are double, and they are situated at points congruent to $\frac{1}{2}$ and $i\Lambda$, respectively.

When substituting in (4-18) we separate the factor $n = 0$ and change negative n to positive. After a slight reordering of the factors one finds

$$F(0) = -4q \prod_{n=1}^{\infty} \left(\frac{1 + q^{2n}}{1 - q^{2n-1}}\right)^4$$

$$F\left(\frac{\omega_1 + \omega_2}{2}\right) = \frac{1}{4} \prod_{n=1}^{\infty} \left(\frac{1 - q^{2n-1}}{1 + q^{2n}}\right)^4,$$

and finally
$$R = \frac{1}{16q} \prod_{n=1}^{\infty} \left(\frac{1 - q^{2n-1}}{1 + q^{2n}} \right)^{8}. \tag{4-19}$$

A similar computation leads to

$$R + 1 = \frac{e_1 - e_2}{e_3 - e_2} = \frac{1}{16q} \prod_{n=1}^{\infty} \left(\frac{1 + q^{2n-1}}{1 + q^{2n}} \right)^{8}. \tag{4-20}$$

In particular, (4-19) and (4-20) give the double inequality

$$16R \leq e^{2\pi\Lambda(R)} \leq 16(R + 1). \tag{4-21}$$

This is a good inequality only when R is large. For small R it should be combined with the identity $\Lambda(R)\Lambda(R^{-1}) = \frac{1}{4}$.

4-13 A DISTORTION THEOREM

The integral inequality (4-6) becomes much more useful when combined with estimates similar to the ones derived in the preceding section. The distortion theorem that we are referring to was originally proved in Ahlfors' thesis [2] by use of the area-length principle together with some differential inequalities. It was actually for the purpose of simplifying this proof that Teichmüller [65] proved Theorem 4-7.

We return to the situation discussed in Sec. 4-5, but this time we regard Ω as part of a strip that may extend to infinity in both directions (Fig. 4-9a). Let the whole strip be mapped conformally on a parallel strip of width 1. Figure 4-9b shows the images E_1', E_2' of E_1, E_2, and the meaning of the numbers α and β as a maximum and minimum. The problem is to find a lower bound for $\beta - \alpha$.

Since we already possess a best possible lower bound for the extremal

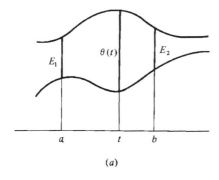

(a)

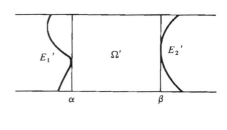

(b)

FIGURE 4-9

distance $d(E_1, E_2)$, and since $d(E_1, E_2) = d(E_1', E_2')$ by conformal invariance, what we need is an upper bound for $d(E_1', E_2')$ in terms of $\beta - \alpha$. An exponential would map the parallel strip on the whole plane, but the disadvantage is that E_1' and E_2' would not be mapped on closed curves. Teichmüller overcomes this difficulty by the simple device of first reflecting the parallel strip across one of the boundary lines, for instance, the real axis. The union of Ω' and its reflected image lies between two symmetric curves $\hat{E}_1'$, $\hat{E}_2'$ whose extremal distance in the double strip is $\frac{1}{2}d(E_1', E_2')$. The exponential $e^{\pi z}$ maps E_1' and E_2' on closed curves C_1, C_2 as shown in Fig. 4-10. Although they appear originally as opposite sides in a quadrilateral, they may also be regarded as contours of an annulus, and because of the symmetry their extremal distance is the same in both cases. We conclude from these considerations that $d(C_1, C_2) = \frac{1}{2}d(E_1, E_2)$.

Now we apply Theorem 4-7. The curve C_1 encloses the origin and passes through a point at distance $e^{\pi \alpha}$ from the origin. C_2 separates C_1 from ∞ and contains a point with absolute value $e^{\pi \beta}$. Except for a normalization this agrees with the situation in the theorem. We obtain $d(C_1, C_2) \leq \Lambda(e^{\pi(\beta - \alpha)})$ and have thereby proved Theorem 4-8.

Theorem 4-8 The mapping illustrated in Fig. 4-9 satisfies

$$\int_a^b \frac{dx}{\theta(x)} \leq 2\Lambda(e^{\pi(\beta - \alpha)}). \tag{4-22}$$

This inequality is the best possible, equality occurring when Ω is a rectangle and E_1', E_2' are half lines extending to infinity in opposite direc-

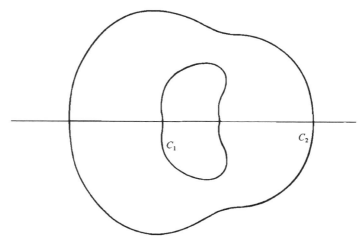

FIGURE 4-10

tions, one on the upper and one on the lower boundary of the parallel strip. For practical purposes it is better to replace (4-22) by a slightly weaker consequence:

Corollary   If $\int_a^b dx/\theta(x) \geq \frac{1}{2}$, then

$$\beta - \alpha \geq \int_a^b \frac{dx}{\theta(x)} - \frac{1}{\pi} \log 32. \tag{4-23}$$

We need only recall that $\Lambda(1) = \frac{1}{2}$, so that the assumption together with (4-22) implies $e^{\pi(\beta-\alpha)} \geq 1$. But (4-21) gives $\Lambda(R) \leq (1/2\pi) \log 32R$ if $R \geq 1$, and (4-23) follows.

4-14   REDUCED EXTREMAL DISTANCE

The extremal distance between two sets will tend to ∞ if one of the sets shrinks to a point. It may happen, however, that the difference between two extremal distances tends to a finite limit.

We assume again that Ω is bounded by a finite number of analytic curves. Let E be the union of a finite number of closed arcs on the boundary. For a fixed $z_0 \in \Omega$ we denote by C_r the circle and by Δ_r the disk with center z_0 and radius r. For $\bar{\Delta}_r \subset \Omega$ we let $d(C_r, E)$ be the extremal distance with respect to $\Omega - \bar{\Delta}_r$. If $r' > r$, the composition law (4-2) yields $d(C_r, E) \geq d(C_{r'}, E) + (1/2\pi) \log (r'/r)$, for the second term is the extremal distance between the two circles. This inequality shows that $d(C_r, E) + (1/2\pi) \log r$ is a decreasing function of r. Hence $\lim_{r\to 0} [d(C_r, E) + (1/2\pi) \log r]$ exists, and under our assumptions it is finite, for if $\Omega \subset \Delta_R$ then $d(C_r, E) \leq (1/2\pi) \log (R/r)$. We denote this limit tentatively by $d(z_0, E)$. This number could serve the same purpose as an extremal distance, except for two drawbacks. First, it need not be positive, and second, it is not a conformal invariant. Both disadvantages are removed by forming the quantity

$$\delta(z_0, E) = d(z_0, E) - d(z_0, C) \tag{4-24}$$

where C is the whole boundary. That $\delta(z_0, E) \geq 0$ is immediate from the comparison principle. That it is also a conformal invariant will emerge from the discussion that follows.

We shall call $\delta(z_0, E)$ the *reduced extremal distance* between z_0 and E. We wish to relate it to other invariant quantities. For this purpose we solve a mixed Dirichlet-Neumann problem to obtain a harmonic function $G(z, z_0)$ with a logarithmic pole at z_0, which is zero on E and satisfies $\partial G/\partial n = 0$ on the rest of the boundary. $G(z, z_0)$ exists and is unique if we add the condition that it be bounded outside a neighborhood of z_0. The behavior at z_0 is of the form

$$G(z,z_0) = -\log|z - z_0| + \gamma(E) + \epsilon(z),$$

where $\gamma(E)$ is a constant and $\epsilon(z) \to 0$ as $z \to z_0$. In the special case in which E is the whole boundary $G(z,z_0)$ is the ordinary Green's function $g(z,z_0)$, and $\gamma(C)$ is the Robin constant relative to z_0.

We need to estimate $d(C_r,E)$ for small r. Let α denote the maximum and β the minimum of $G(z,z_0)$ when $z \in C_r$. Then the level curve $L_\alpha = \{z; G(z,z_0) = \alpha\}$ will lie inside C_r while L_β lies outside C_r. The comparison principle yields

$$d(L_\beta,E) \leq d(C_r,E) \leq d(L_\alpha,E).$$

But we know by Theorem 4-5 that $d(L_\alpha,E) = 1/D(u)$, where $u = G/\alpha$ and the Dirichlet integral is extended over the region between L_α and C. Since $D(G/\alpha) = \alpha^{-2}D(G)$ and

$$D(G) = \int_{L_\alpha} G \frac{\partial G}{\partial n}|dz| = \alpha \int_{L_\alpha} \frac{\partial G}{\partial n}|dz| = 2\pi\alpha$$

we obtain $d(C_r,E) \leq \alpha$, and similarly $d(C_r,E) \geq \beta/2\pi$.

From the development of $G(z,z_0)$ it follows that α and β are both of the form $-\log r + \gamma(E) + \epsilon(r)$. Hence $d(C_r,E) = (1/2\pi)[-\log r + \gamma(E) + \epsilon(r)]$ and we conclude by (4-24) that

$$\delta(z_0,E) = \frac{1}{2\pi}[\gamma(E) - \gamma(C)]. \tag{4-25}$$

Incidentally, this proves that $\gamma(C) \leq \gamma(E)$. Also, the way $G(z,z_0)$ and $g(z,z_0)$ change under conformal mapping makes it clear that the right-hand side in (4-26) is a conformal invariant.

The result gains in significance if we show, as we shall, that $\delta(z_0,E)$ solves an extremal problem which is similar to, but still fundamentally different from, the problem of extremal distance.

Theorem 4-9 The number $1/\delta(z_0,E)$ is the minimum of the Dirichlet integral $D(v)$ in the class of functions v with the following properties:

    i) v is subharmonic and of class C^1 in Ω;
    ii) v has a continuous extension to C;
    iii) $v(z) \leq 0$ on E and $v(z_0) \geq 1$.

The similarity to extremal distance is seen by considering $\rho = |\text{grad } v|$. Indeed, we have $\int_\gamma \rho|dz| \geq 1$ for all arcs γ that join z_0 and E. For the family Γ of such arcs we thus have $L(\Gamma,\rho) \geq 1$ while $A(\Omega,\rho) = D(v)$.

Therefore while $L(\Gamma,\rho)^2/A(\Omega,\rho)$ becomes arbitrarily large when ρ varies over all metrics, the theorem tells us that the maximum is $\delta(z_0,E)$ when ρ is restricted to be the gradient of a subharmonic function.

For the proof we compare v with the function $G - g$. If v has a finite Dirichlet integral, it is not difficult to show the validity of the formula

$$D(v,G - g) = \int_C v \frac{\partial(G - g)}{\partial n} |dz| = \int_E v \frac{\partial G}{\partial n} |dz| - \int_C v \frac{\partial g}{\partial n} |dz|. \quad (4\text{-}26)$$

The first integral on the right has a nonnegative integrand, for $v \leq 0$ and $\partial G/\partial n < 0$, because $G > 0$ inside the region and $G = 0$ on E. The second integral is a generalized Poisson integral and is equal to 2π times the value at z_0 of the harmonic function u with the same boundary values as v. Because v is subharmonic, we have $v(z_0) \leq u(z_0)$, so that

$$- \int_C v \frac{\partial g(z,z_0)}{\partial n} |dz| \geq 2\pi v(z_0) \geq 2\pi.$$

We conclude that $D(v,G - g) \geq 2\pi$.

On the other hand, $G - g$ is harmonic throughout Ω, and we obtain

$$D(G - g) = \int_C (G - g) \frac{\partial(G - g)}{\partial n} |dz|$$

$$= - \int_C (G - g) \frac{\partial g}{\partial n} |dz| = 2\pi[\gamma(E) - \gamma(C)]$$

because $\gamma(E) - \gamma(C)$ is the value of $G - g$ at z_0. Now the Schwarz inequality gives $4\pi^2 \leq D(v,G - g)^2 \leq D(v)D(G - g) = 2\pi[\gamma(E) - \gamma(C)]D(v)$, and in view of (4-25) we have shown that $1/D(v) \leq \delta(z_0,E)$. The upper bound is reached, namely, for a constant multiple of $G - g$. The function

$$v(z) = \frac{G(z,z_0) - g(z,z_0)}{\gamma(E) - \gamma(C)}$$

does in fact satisfy conditions (i) through (iii), and $D(v) = 1/\delta(z_0,E)$.

Theorem 4-8 should be compared with Theorem 2-4. The latter is a special case and implies that the reduced extremal distance between the origin and a set E on the unit circle is equal to $-(1/\pi) \log E$.

EXERCISES

1 For any arc γ let $\bar{\gamma}$ denote its reflection in the real axis and let γ^+ be obtained by reflecting the part below the real axis and retaining the part above it. The notations $\bar{\Gamma}$ and Γ^+ are self-explanatory. If $\Gamma = \bar{\Gamma}$, show that $\lambda(\Gamma) = \frac{1}{2}\lambda(\Gamma^+)$.

2   Find the maximum extremal distance between an arc on a circle and a continuum that joins the center to the circumference. Show that its value can be expressed in terms of the function Λ.

3   Let the sides of a triangle be numbered. Find $\lambda(\Gamma)$ when Γ consists of all arcs that begin on side 1, touch side 2, and end on side 3.

Ans. $\lambda(\Gamma) = 2$.

4   Given two points a and b, let Γ be the family of figure-eight-shaped curves with winding number 1 about a and -1 about b. Show that $\lambda(\Gamma) = 4$. (Make use of the preceding exercise. The corresponding question for an arbitrary triply connected region is open.)

5   Find the extremal length of the nondividing closed curves on a Möbius band. (Use the remark in the last paragraph of Sec. 4-4.)

NOTES   The articles by Grötzsch are from the period 1928–1934, and they were all published in *Verhandlungen der sächsischen Akademie der Wissenschaften, Leipzig.* Because of the relative obscurity of this journal, it was a long time before Grötzsch's work became generally known.

The definition and underlying idea of extremal length were first conceived by Beurling, presumably in 1943–1944. They were first made public at the Scandinavian Congress of Mathematicians in Copenhagen, 1946, in parallel papers read by Beurling and by Ahlfors [1]; Beurling's paper never appeared in print. The first systematic account was given in a joint article in publications of the Bureau of Standards in 1949, but the best known version is in Ahlfors' and Beurling's article on function theoretic null sets [2].

Because of this history Ahlfors has sometimes been given partial credit for the discovery of extremal length. A more justifiable claim would be that of codeveloper.

An acccunt based on Ahlfors' lectures at Harvard University and in Japan in 1957 has been published in Ohtsuka's recent book [48].

Theorems 4-5 and 4-6 are almost certainly due to Beurling.

The connection between prime ends and extremal length was explored by E. Schlesinger [58].

5

ELEMENTARY THEORY OF UNIVALENT FUNCTIONS

5-1 THE AREA THEOREM

An analytic or meromorphic function in Ω is said to be *univalent* or *schlicht* if $f(z_1) = f(z_2)$ only when $z_1 = z_2$. We shall deal only with the case in which Ω is simply connected. In fact, we assume Ω to be a circular region. There are two standard normalizations:

1) $f \in S$ if f is univalent and holomorphic in $|z| < 1$ with the development

$$f(z) = z + a_2 z^2 + \cdots + a_n z^n + \cdots.$$

2) $F \in \Sigma$ if F is univalent in $|z| > 1$ with the development

$$F(z) = z + \frac{b_1}{z} + \cdots + \frac{b_n}{z^n} + \cdots.$$

The famous coefficient problem is to find necessary and sufficient conditions for a_n and b_n. Bieberbach's conjecture $|a_n| \leq n$ has been proved for $n = 2,3,4,5,6$.

The following theorem is known as the *area theorem*. It was first proved by Gronwall [23] and rediscovered by Bieberbach [7].

Theorem 5-1   All $F \in \Sigma$ satisfy $\sum_{1}^{\infty} n|b_n|^2 \le 1$.

PROOF   Let C_ρ be the circle $|z| = \rho > 1$ with positive orientation, and set

$$I_\rho(F) = \frac{i}{2} \int_{C_\rho} F \, d\bar{F}.$$

If $F = u + iv$ and if Γ_ρ denotes the image curve of C_ρ, we have

$$I_\rho(F) = \int_{\Gamma_\rho} u \, dv,$$

and by elementary calculus this represents the area enclosed by Γ_ρ. Hence $I_\rho(F) > 0$.

Direct calculation gives

$$I_\rho(F) = \frac{i}{2} \int_{C_\rho} \left(z + \sum_{1}^{\infty} b_n z^{-n} \right) \left(1 - \sum_{1}^{\infty} n \bar{b}_n \bar{z}^{-n-1} \right) d\bar{z}$$

$$= \frac{1}{2} \int_{C_\rho} \left(z + \sum_{1}^{\infty} b_n z^{-n} \right) \left(\bar{z} - \sum_{1}^{\infty} n \bar{b}_n \bar{z}^{-n} \right) d\theta = \pi \left[\rho^2 - \sum_{1}^{\infty} n|b_n|^2 \rho^{-2n} \right].$$

Thus $\sum_{1}^{\infty} n|b_n|^2 \rho^{-2n} < \rho^2$, and the theorem follows for $\rho \to 1$.

A particular consequence is $|b_1| \le 1$, and this is sharp, for $z + e^{i\beta}/z$ is *schlicht*. It is clearly the only case of equality.

Theorem 5-2   All $f \in S$ satisfy $|a_2| \le 2$ with equality only for the so-called Koebe functions $f(z) = z(1 + e^{i\beta}z)^{-2}$.

PROOF   An obvious attempt is to pass from $f \in S$ to $F(z) = f(z^{-1})^{-1} + a_2 \in \Sigma$ whose development begins with $F(z) = z + (a_2{}^2 - a_3)z^{-1} + \cdots$. Theorem 5-1 gives $|a_2{}^2 - a_3| \le 1$, which is interesting, but not what we want.

We employ a famous device due to Faber. Because $f(z)/z$ is holomorphic and $\ne 0$, we can define $h(z) = [f(z)/z]^{\frac{1}{2}}$, $h(0) = 1$, and subsequently $g(z) = zh(z^2)^{\frac{1}{2}}$. The function g is univalent, for $g(z_1) = g(z_2)$ implies $f(z_1{}^2) = f(z_2{}^2)$, hence $z_1 = z_2$ or $z_1 = -z_2$; the latter is ruled out because g is odd and $\neq 0$ for $z \neq 0$. The development of g is $g(z) = z + \frac{1}{2}a_2z^3 + \cdots$. The preceding result yields $|a_2| \leq 2$.

For equality we must have $|b_1| = 1$ in the area theorem, hence $F(z) = g(z^{-1})^{-1} = z + e^{i\beta}z^{-1}$, which leads to $f(z) = z(1 + e^{i\beta}z)^{-2}$. The Koebe function maps the unit disk on the complement of a slit $\{|w| \geq \frac{1}{4},\ \arg w = -\beta\}$. The coefficients of the Koebe function satisfy $|a_n| = n$.

The inequality $|a_2| \leq 2$ leads immediately to upper and lower bounds for $|f(z)|$ and $|f'(z)|$. These estimates are collectively known as the *distortion theorem*.

Theorem 5-3   The functions $f \in S$ satisfy

$$|z|(1 + |z|)^{-2} \leq |f(z)| \leq |z|(1 - |z|)^{-2} \tag{5-1}$$
$$(1 - |z|)(1 + |z|)^{-3} \leq |f'(z)| \leq (1 + |z|)(1 - |z|)^{-3}, \tag{5-2}$$

with equality only for the Koebe functions.

PROOF   Consider $f(Tz)$, where T is a conformal mapping of the unit disk onto itself. The function $f \circ T$ is again *schlicht*, but not normalized. We have

$$(f \circ T)' = (f' \circ T)T''$$
$$(f \circ T)'' = (f'' \circ T)T'^2 + (f' \circ T)T'',$$

and the Taylor development at 0 is

$$f \circ T = f(T0) + f'(T0)T'(0)z$$
$$+ \frac{1}{2}[f''(T0)T'(0)^2 + f'(T0)T''(0)]z^2 + \cdots.$$

Theorem 5-2 yields

$$\left| \frac{f''(T0)}{f'(T0)}T'(0) + \frac{T''(0)}{T'(0)} \right| \leq 4.$$

We choose $Tz = (z + \zeta)(1 + \bar{\zeta}z)^{-1}$ with $|\zeta| < 1$. Then $T0 = \zeta$, $T'(0) = 1 - |\zeta|^2$, $T''(0)/T'(0) = -2\bar{\zeta}$. The inequality becomes

$$\left| \frac{f''(\zeta)}{f'(\zeta)} - \frac{2\bar{\zeta}}{1 - |\zeta|^2} \right| \leq \frac{4}{1 - |\zeta|^2}. \tag{5-3}$$

Define $\log f'(z)$ so that $\log f'(0) = 0$. Integration of (5-3) along a radius yields

$$\left| \log f'(z) - \int_0^{|z|} \frac{2r\,dr}{1-r^2} \right| \leq \int_0^{|z|} \frac{4\,dr}{1-r^2},$$

and hence

$$\log \frac{1}{1-|z|^2} - 2\log \frac{1+|z|}{1-|z|} \leq \log |f'(z)| \leq \log \frac{1}{1-|z|^2} + 2\log \frac{1+|z|}{1-|z|}.$$

This is the double inequality (5-2).

From (5-2) we have at once

$$|f(z)| \leq \int_0^{|z|} \frac{1+r}{(1-r)^3}\,dr = \frac{|z|}{(1-|z|)^2},$$

which is the upper bound in (5-1). To find the lower bound, let $m(r)$ denote the minimum of $|f(z)|$ on $|z| = r$. The image of $|z| < r$ contains the disk $|w| < m(r)$. Therefore, there exists a curve γ from 0 to $|z| = r$ such that

$$m(r) = \int_\gamma |f'(z)|\,|dz|.$$

Since γ intersects all circles $|z| = \rho < r$, the lower bound for $|f'(z)|$ leads to the desired estimate

$$m(r) \geq \int_0^r \frac{1-\rho}{(1+\rho)^3}\,d\rho = \frac{r}{(1+r)^2}.$$

For equality to hold it is necessary to have equality in (5-3) on the radius from 0 to z, and hence in particular at 0. This means that $|a_2| = 2$, and f must be a Koebe function.

As $r \to 1$ the lower bound for $|f(z)|$ tends to $\frac{1}{4}$.

Corollary   The image of the unit disk under a mapping $f \in S$ contains the disk with center 0 and radius $\frac{1}{4}$.

This is Koebe's one-quarter theorem, which we already proved in Sec. 2-3 as an application of capacity. Koebe did not give the value of the constant.

5-2   THE GRUNSKY AND GOLUSIN INEQUALITIES

A function is said to be m-valent if it assumes each value at most m times. Theorem 5-1 can be generalized to this situation.

Theorem 5-4 If F is analytic and m-valent for $|z| > 1$ with a development

$$F(z) = \sum_{-m}^{\infty} b_n z^{-n} \qquad b_{-m} \neq 0,$$

then
$$\sum_{1}^{\infty} n |b_n|^2 \leq \sum_{1}^{m} n |b_{-n}|^2. \qquad (5\text{-}4)$$

For the proof we need to know that $I_\rho(F)$ is still positive. This fact is not as obvious as before, and we prove it as a separate lemma.

Lemma 5-1 If F is m-valent for $|z| > 1$ with a pole of order m at ∞, then

$$I_\rho(F) = \frac{i}{2} \int_{|z|=\rho} F \, d\bar{F} > 0.$$

PROOF Given $w \neq \infty$ let $n(w)$ be the number of roots of $F(z) = w$ in $|z| > \rho$. Assuming that $F \neq w$ on C_ρ we have, for large R,

$$n(w) = \frac{1}{2\pi i} \int_{C_R - C_\rho} \frac{dF}{F - w} = m - \frac{1}{2\pi i} \int_{C_\rho} \frac{dF}{F - w},$$

and hence
$$\frac{1}{2\pi i} \int_{C_\rho} \frac{dF}{F - w} \geq 0. \qquad (5\text{-}5)$$

Choose M greater than the maximum of $|F(z)|$ on C_ρ. We integrate (5-5) with respect to w over the disk $|w| < M$. The triple integral is absolutely convergent, and since the image of C_ρ covers only a null set, we obtain

$$\frac{1}{2\pi i} \int_{C_\rho} \left(\iint_{|w|<M} \frac{du \, dv}{F - w} \right) dF \geq 0 \qquad (w = u + iv). \qquad (5\text{-}6)$$

The double integral is calculated in standard fashion:

$$\iint_{|w|<M} \frac{du \, dv}{F - w} = -\frac{i}{2} \iint_{|w|<M} \frac{d\bar{w} \, dw}{F - w}$$

$$= -\frac{i}{2} \int_{|w|=M} \frac{\bar{w} \, dw}{F - w} + \lim_{\epsilon \to 0} \frac{i}{2} \int_{|w-F|=\epsilon} \frac{\bar{w} \, dw}{F - w}.$$

The first integral vanishes as seen by setting $\bar{w} = M^2/w$, and the second has the limit $\pi\bar{F}$. Hence

$$\iint\limits_{|w|<M} \frac{du\,dv}{F-w} = \pi\bar{F},$$

and substitution in (5-6) yields $I_\rho(F) \geq 0$.

The theorem follows from the lemma by explicit calculation of $I_\rho(F)$, exactly as in the proof of Theorem 5-1. In our new notation we find

$$\sum_{-m}^{\infty} n|b_n|^2 \leq 0.$$

Separation of the positive and negative terms yields Theorem 5-4.

Corollary   With the hypothesis and notation of Theorem 5-4 it is also true that

$$\sum_{1}^{m} n|b_n|\,|b_{-n}| \leq \sum_{1}^{m} n|b_{-n}|^2. \tag{5-7}$$

The theory of m-valent functions is less interesting than that of univalent functions. For this reason the main importance of Theorem 5-4 is to serve as a tool for the study of univalent functions. Let P_m be an arbitrary polynomial of degree m. If $F \in \Sigma$ (univalent with pole at ∞), then $P_m(F)$ is obviously m-valent with a pole of order m at ∞. Therefore, the coefficients of $P_m(F)$ satisfy (5-4) and (5-7). In this way we obtain a great deal of information about F, albeit in rather implicit form. In the exercise section of this chapter we shall show how to translate this information into explicit inequalities. The inequalities that arise from (5-4) are known as the Golusin inequalities. The Grunsky inequalities, which were discovered earlier, are easy consequences of (5-7).

5-3   PROOF OF $|a_4| \leq 4$

The inequality $|a_4| \leq 4$ was first proved by Garabedian and Schiffer [21]. Later Charzynski and Schiffer [14] made the important discovery that $|a_4| \leq 4$ can be proved directly, and with much less work, from the Grunsky inequalities. We shall follow their lead, but we shall use (5-4) rather than (5-7).

Let

$$F(z) = z + b_1z^{-1} + b_3z^{-3} + b_5z^{-5} + \cdots \tag{5-8}$$

be an odd univalent function in $|z| > 1$, and set

$$F(z)^3 = z^3 + c_{-1}z + c_1z^{-1} + c_3z^{-3} + \cdots.$$

The coefficients are

$$
\begin{aligned}
c_{-1} &= 3b_1 \\
c_1 &= 3b_1{}^2 + 3b_3 \\
c_3 &= b_1{}^3 + 6b_1b_3 + 3b_5.
\end{aligned}
\tag{5-9}
$$

We introduce a complex parameter t and apply Theorem 5-4 to the function $F(z)^3 + tF(z)$ which is 3-valent with a triple pole. This gives

$$
|tb_1 + c_1|^2 + 3|tb_3 + c_3|^2 \le |t + c_{-1}|^2 + 3,
\tag{5-10}
$$

and after rearrangement,

$$
(1 - |b_1|^2 - 3|b_3|^2)|t|^2 + 2 \operatorname{Re} t(\bar{c}_{-1} - b_1\bar{c}_1 - 3b_3\bar{c}_3) \\
+ 3 + |c_{-1}|^2 - |c_1|^2 - 3|c_3|^2 \ge 0.
$$

We already know from the area theorem that the coefficient of $|t|^2$ is nonnegative. In addition, the positive definiteness of the Hermitian form implies

$$
|\bar{c}_{-1} - b_1\bar{c}_1 - 3b_3\bar{c}_3|^2 \\
\le (1 - |b_1|^2 - 3|b_3|^2)(3 + |c_{-1}|^2 - |c_1|^3 - 3|c_3|^2).
\tag{5-11}
$$

We conclude that c_3 lies inside a certain circle:

$$
|c_3 - \omega| \le \rho.
\tag{5-12}
$$

To find explicit expressions for ω and ρ we bring (5-11) to the form

$$
\left| c_3 - \frac{b_3(c_{-1} - \bar{b}_1c_1)}{1 - |b_1|^2} \right|^2 \le \frac{(1 - |b_1|^2 - 3|b_3|^2)(3 + |c_{-1}|^2 - |c_1|^2)}{3(1 - |b_1|^2)^2} \\
- \frac{|c_{-1} - \bar{b}_1c_1|^2}{3(1 - |b_1|^2)} + \frac{|b_3|^2|c_{-1} - \bar{b}_1c_1|^2}{(1 - |b_1|^2)^2}.
$$

With the aid of (5-9) we thus have

$$
\omega = 3b_1b_3 - \frac{3\bar{b}_1b_3{}^2}{1 - |b_1|^2},
\tag{5-13}
$$

and on using the identity

$$
|c_{-1} - \bar{b}_1c_1|^2 - |b_1c_{-1} - c_1|^2 = (|c_{-1}|^2 - |c_1|^2)(1 - |b_1|^2),
$$

it turns out that ρ has the surprisingly simple value

$$
\rho = \frac{1 - |b_1|^2 - 3|b_3|^2}{1 - |b_1|^2}.
\tag{5-14}
$$

Now let $f(z) = z + a_2z^2 + \cdots$ be univalent in $|z| < 1$ and form $F(z) = f(z^{-2})^{-\frac{1}{2}}$. Then F is univalent of the form (5-8), and the relations between the coefficients are

$$a_2 = -2b_1$$
$$a_3 = -2b_3 + 3b_1{}^2$$
$$a_4 = -2b_5 + 6b_1b_3 - 4b_1{}^3.$$

From these relations together with (5-9) and (5-13) we obtain

$$c_3 - \omega = -\tfrac{3}{2}a_4 - 5b_1{}^3 + 12b_1b_3 + \frac{3\bar{b}_1b_3{}^2}{1 - |b_1|^2},$$

and by (5-13) and (5-14)

$$|a_4| \le \left| \frac{10}{3}b_1{}^3 - 8b_1b_3 - \frac{2\bar{b}_1b_3{}^2}{1 - |b_1|^2} \right| + \frac{2}{3} - \frac{2|b_3|^2}{1 - |b_1|^2}. \qquad (5\text{-}15)$$

If $b_1 = 0$, there is nothing left to prove, and if $b_1 \ne 0$, we can write $b_3 = sb_1{}^2$. With this notation (5-15) becomes

$$|a_4| \le \frac{2|b_1|^5}{1 - |b_1|^2} \left| s^2 + 4\left(s - \frac{5}{3}\right) \frac{1 - |b_1|^2}{|b_1|^2} \right| + \frac{2}{3} - \frac{2|s|^2|b_1|^4}{1 - |b_1|^2}. \qquad (5\text{-}16)$$

We need an estimate for the absolute value of the quadratic polynomial in the first term on the right. With simpler notations we shall show that

$$|s^2 + 2\alpha s - \beta|^2 \le 1 + \frac{\alpha^2}{\beta}(|s|^2 + \beta)^2, \qquad (5\text{-}17)$$

provided that α is real and $\beta > 0$. To see that this is so we set $\operatorname{Re} s = u$ and observe that $\operatorname{Re} s^2 = 2u^2 - |s|^2$. We obtain

$$|s^2 + 2\alpha s - \beta|^2 = |s|^4 + (4\alpha^2 + 2\beta)|s|^2 + \beta^2 + 4\alpha(|s|^2 - \beta)u - 4\beta u^2,$$

and (5-17) follows on replacing the right-hand side by its maximum for variable u and fixed $|s|$.

We apply (5-17) to (5-16) and find

$$|a_4| \le \frac{2|b_1|^2}{1 - |b_1|^2} \left(\frac{12 - 7|b_1|^2}{5} \right)^{\frac{1}{2}} (|s|^2|b_1|^2 + \tfrac{5}{3} - \tfrac{5}{3}|b_1|^2) + \frac{2}{3} - \frac{2|s|^2|b_1|^4}{1 - |b_1|^2}.$$

In this formula the expression on the right increases with $|s|$ because $|b_1| \le 1$. We replace $|s|^2$ by its upper bound $\frac{1}{3}(1 - |b_1|^2)|b_1|^{-4}$ and find

$$|a_4| \le \frac{2 + 10|b_1|^2}{3} \left(\frac{12 - 7|b_1|^2}{5} \right)^{\frac{1}{2}}.$$

On setting $1 - |b_1|^2 = \lambda$, the final inequality reads

$$|a_4|^2 \leq \tfrac{4}{45}(6 - 5\lambda)^2(5 + 7\lambda) = 16 - \tfrac{64}{15}\lambda - \tfrac{4}{9}\lambda^2(59 - 35\lambda) \leq 16.$$

We have proved a little more than $|a_4| \leq 4$, for instance,

$$16 - |a_4|^2 \geq \tfrac{16}{15}(4 - |a_2|^2). \tag{5-18}$$

NOTES   The passage from the area theorem via $|a_2| \leq 2$ to the distortion theorem was first pointed out by R. Nevanlinna [47]. It has been generalized to classes with $|a_2| \leq \alpha$.

The article with the Grunsky inequalities is Grunsky [25]; actually, it covers a much more general situation. The best reference to Golusin is his textbook (Golusin [22]). Golusin's method, with some variations, can also be found in Jenkins [33] and Pommerenke [52, 53]. An excellent account on Faber polynomials is Curtiss [16].

The proof of (5-18) has appeared in a Russian publication honoring M. A. Lavrentiev.

EXERCISES

1   If $F(z) = z + \sum_1^\infty b_n z^{-n}$ is analytic for $|z| > 1$, show that

$$\frac{F'(\zeta)}{F(\zeta) - w} = \sum_0^\infty P_m(w)\zeta^{-m-1},$$

where the P_m are polynomials of degree m with leading coefficient 1. The development is valid in a neighborhood of $\zeta = \infty$, which depends on w.

The P_m are known as the Faber polynomials associated with F.

2   Under the same conditions, prove carefully that for a suitable branch of the logarithm there is a development of the form

$$\log \frac{F(\zeta) - F(z)}{\zeta - z} = -\sum_{m=1}^\infty \sum_{n=1}^\infty b_{mn}\zeta^{-m}z^{-n}$$

valid for sufficiently large $|\zeta|$ and $|z|$.

Deduce that

$$P_m[F(z)] = z^m + m\sum_{n=1}^\infty b_{mn}z^{-n}.$$

Prove further that P_m is the only polynomial P such that the singular part of $P[F(z)]$ at ∞ reduces to z^m.

3 Assume F to be univalent. Apply Theorem 5-4 to $P = \sum_{k=1}^{m} k^{-1} l_k P_k$ with arbitrary complex t_k to obtain

$$\sum_{n=1}^{\infty} n \left| \sum_{k=1}^{\infty} b_{kn} t_k \right|^2 \leq \sum_{k=1}^{\infty} k^{-1} |t_k|^2 \qquad \text{(Golusin)}$$

$$\left| \operatorname{Re} \sum_{k,n=1}^{\infty} b_{kn} t_k t_n \right| \leq \sum_{k=1}^{\infty} k^{-1} |t_k|^2 \qquad \text{(Grunsky)}.$$

4 Show conversely that F is univalent if all Golusin or Grunsky inequalities are fulfilled by proving that the series in Exercise 2 converges for all $|\zeta| > 1$, $|z| > 1$.

6

LÖWNER'S METHOD

6-1 APPROXIMATION BY SLIT MAPPINGS

It was a remarkable feat of Löwner to prove $|a_3| \le 3$ in 1923 when
nothing was known about univalent functions beyond the most elementary
results. It seems to be an experimental fact that the Grunsky inequalities
are of no use for odd coefficients. Löwner's method is quite simple, but
slightly delicate. Its usefulness goes beyond proving coefficient inequalities.

We may assume that $f(z)$ is univalent and analytic in the closed unit
disk. For the moment we abandon the normalization $f'(0) = 1$ and require
instead that $|f(z)| < 1$ for $|z| \le 1$, $f(0) = 0$, and $f'(0) > 0$. The image
B is thus contained in the unit disk. We approximate B by a one-parameter
family of simply connected subregions of the unit disk obtained by omit-
ting a slit which begins on the unit circle, leads to a boundary point of B,
and then follows the boundary, stopping short of the initial point of
contact (Fig. 6-1). The equation of the slit is written as $w = \gamma(t)$, $0 \le
t < t_0$, where γ is continuous and one to one. The region B_t is the comple-
ment of the arc $\gamma[0,t]$ with respect to the disk. The slit closes up for $t = t_0$
so that $B_{t_0} = B$. Note the discontinuous change in B_t when $t \to t_0$.

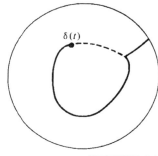

FIGURE 6-1

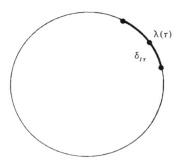

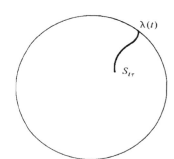

FIGURE 6-2

Let $f_t(z)$ be the Riemann mapping function from $|z| < 1$ to B_t with $f_t(0) = 0$, $f_t'(0) > 0$. We know from the theory of boundary correspondence (Sec. 4-6) that f_t has a continuous extension to $|z| \leq 1$, and that there is a unique point $\lambda(t)$ on $|z| = 1$ with $f_t[\lambda(t)] = \gamma(t)$, the tip of the slit.

We proceed to establish some continuity properties. Consider two parameter values t and τ with $t < \tau < t_0$. The function $h_{t\tau} = f_t^{-1} \circ f_\tau$ is defined and univalent in $|z| < 1$, and it has a continuous extension to the closed disk. The reader should convince himself that the mapping by $h_{t\tau}$ is as indicated in Fig. 6-2. In words, the unit circle is mapped on itself, except for an arc $\delta_{t\tau} = f^{-1}(\gamma[t,\tau])$ which is mapped on a slit $S_{t\tau} = f_t^{-1}(\gamma[t,\tau])$. If τ is fixed and t increases to τ, it is immediately clear that $\delta_{t\tau}$ closes down on the point $\lambda(\tau)$. Similarly, if t is fixed and τ decreases to t, the slit $S_{t\tau}$ will be shortened until it shrinks to $\lambda(t)$.

The case $\tau = t_0$ needs separate consideration. It is seen that

$$\delta_{t t_0} = f_{t_0}^{-1}(\gamma[t,t_0])$$

is still an arc on the unit circle which shrinks to its end point $\lambda(t_0)$ as

$t \nearrow t_0$. Its image S_{tt_0} is no longer a slit, but an arc with one end point at $\lambda(t)$.

Lemma 6-1 $|f_t^{-1}(w)|$ is strictly increasing and $f_t'(0)$ is strictly decreasing as functions of t.

PROOF   This is immediate by Schwarz's lemma. Indeed, since $|h_{tr}(z)| < 1$ for $|z| < 1$ and $h_{tr}(0) = 0$, we have $|h_{tr}(z)| \leq |z|$, which is the same as $|f_t^{-1}(w)| \leq |f_\tau^{-1}(w)|$. For $w = 0$ this becomes $|f_t'(0)| \geq |f_\tau'(0)|$. In both cases equality is excluded because h_{tr} is not the identity mapping.

Lemma 6-2 $f_t'(0)$ and $f_t^{-1}(w)$ are continuous on the left.

Consider the Poisson-Schwarz representation

$$\log \frac{h_{tr}(z)}{z} = \frac{1}{2\pi} \int_0^{2\pi} \frac{e^{i\theta} + z}{e^{i\theta} - z} \log |h_{tr}(e^{i\theta})| \, d\theta. \tag{6-1}$$

The integrand is zero except on δ_{tr}, and on setting $z = f_\tau^{-1}(w)$ we find

$$\log \frac{f_t^{-1}(w)}{f_\tau^{-1}(w)} = \frac{1}{2\pi} \int_{\delta_{tr}} \frac{e^{i\theta} + f_\tau^{-1}(w)}{e^{i\theta} - f_\tau^{-1}(w)} \log |h_{tr}(e^{i\theta})| \, d\theta. \tag{6-2}$$

For $w = 0$ this becomes

$$\log \frac{f_\tau'(0)}{f_t'(0)} = \frac{1}{2\pi} \int_{\delta_{tr}} \log |h_{tr}(e^{i\theta})| \, d\theta. \tag{6-3}$$

We know from Lemma 6-1 that $f_t'(0)$ has a limit $\geq f_\tau'(0)$ as t increases to τ. We denote this limit by $e^\alpha f_\tau'(0)$ so that $\alpha \geq 0$ and, according to 6-2,

$$\lim_{t \nearrow \tau} \frac{1}{2\pi} \int_{\delta_{tr}} \log |h_{tr}| \, d\theta = -\alpha. \tag{6-4}$$

Because δ_{tr} shrinks to a point, comparison of (6-1) or (6-2) with (6-4) shows at once that

$$\lim_{t \nearrow \tau} h_{tr}(z) = z \exp\left[-\alpha \frac{\lambda(\tau) + z}{\lambda(\tau) - z} \right]. \tag{6-5}$$

As a limit of univalent functions the function on the right is either univalent or constant. If $\alpha > 0$, neither is true, for then the function tends to 0 as $z \to \lambda(\tau)$ radially. Hence $\alpha = 0$, and we have proved the left continuity of $f_t'(0)$. At the same time (6-5) becomes

$$\lim_{t \nearrow \tau} h_{tr}(z) = z, \tag{6-6}$$

which is equivalent to $f_t^{-1}(w) \rightarrow f_\tau^{-1}(w)$ for all $w \in B_\tau$. Note that the proof needs no change if $\tau = t_0$.

The proof also shows that (6-6) holds uniformly on every closed subset of $|z| \leq 1$ that does not include $\lambda(\tau)$.

Lemma 6-3 $f_t'(0)$ and $f_t^{-1}(w)$ are continuous on the right.

This time we keep t fixed and let τ decrease to t. Since $S_{t\tau}$ shrinks to a point, $h_{t\tau}^{-1}(\zeta) = f_\tau^{-1}[f_t(\zeta)]$ is ultimately defined in any disk $|\zeta| < 1 - \epsilon$, $0 < \epsilon < 1$. Schwarz's lemma gives $|h_{t\tau}^{-1}(\zeta)| \leq (1 - \epsilon)^{-1}|\zeta|$, or $|f_\tau^{-1}(w)| \leq (1 - \epsilon)^{-1}|f_t^{-1}(w)|$. Since we already know that $|f_t^{-1}| < |f_\tau^{-1}|$, the right continuity of $|f_t^{-1}|$ has been proved. That of f_t^{-1} follows routinely, for instance, by use of the Poisson-Schwarz representation.

Lemma 6-4 $S_{t\tau}$ approaches $\lambda(\tau)$ as $t \nearrow \tau$, and $\delta_{t\tau}$ tends to $\lambda(t)$ as $\tau \searrow t$.

This is best proved by use of the argument principle. Observe first that $h_{t\tau}$ and $h_{t\tau}^{-1}$ can be extended by symmetry across the unit circle, one to the full complement of $\delta_{t\tau}$, the other to the complement of $S_{t\tau}$ and its reflection. Let C_ϵ be a circle with center $\lambda(\tau)$ and small radius ϵ. We have already remarked that $h_{t\tau}(z)$ tends uniformly to z on any closed set that does not contain $\lambda(\tau)$. This is also true for the reflected function, and thus on all of C_ϵ. When t is close to τ, the image of C_ϵ will lie close to C_ϵ, for instance, inside the circle with center $\lambda(\tau)$ which passes through a given point ζ outside C_ϵ. The image curve therefore has a winding number zero about ζ, and it follows that $h_{t\tau} - \zeta$ has as many zeros as poles outside C_ϵ. Since there is a pole at ∞, the function h must assume the value ζ. But no value on $S_{t\tau}$ is assumed, proving that $S_{t\tau}$ lies inside C_ϵ. The reasoning remains valid for $\tau = t_0$.

The second part is proved in the same way by applying the argument principle to $h_{t\tau}^{-1}$. The reader should be aware of a small difficulty. The proof of Lemma 6-3 shows that $h_{t\tau}^{-1}(\zeta) \rightarrow \zeta$ uniformly, but only on compact parts of the open disk, and we need uniformity on C_ϵ. To fill the gap, represent $h_{t\tau}^{-1}$ by its Cauchy integral over $|\zeta| = 2$, and $C_{\epsilon'}$ with $\epsilon' < \epsilon$. The integral over $C_{\epsilon'}$ tends to zero as $\epsilon' \rightarrow 0$, and it becomes clear that $h_{t\tau}^{-1}$ tends uniformly to the identity on any compact set that does not contain $\lambda(t)$, and in particular on C_ϵ.

Lemma 6-5 $\lambda(t)$ is continuous.

This follows from Lemma 6-4 because $\lambda(t) \in S_{t\tau}$ and $\lambda(\tau) \in \delta_{t\tau}$.

6-2 LÖWNER'S DIFFERENTIAL EQUATION

Having proved that $f_t'(0)$ is a continuous strictly decreasing function we are free to choose $-\log f_t'(0)$ as a new parameter. In other words, we may assume that $f_t'(0) = e^{-t}$. With this normalization (6-3) becomes

$$t - \tau = \frac{1}{2\pi} \int_{\delta_{t\tau}} \log |h_{t\tau}(e^{i\theta})| \, d\theta. \tag{6-7}$$

It follows by (6-2) and (6-7) that

$$\frac{\partial}{\partial t} \log f_t^{-1}(w) = \frac{\lambda(t) + f_t^{-1}(w)}{\lambda(t) - f_t^{-1}(w)}. \tag{6-8}$$

Indeed, $f_t^{-1}(w)$ is continuous, and $\delta_{t\tau}$ shrinks to a point, trivially when $t \nearrow \tau$ and by virtue of the second part of Lemma 6-4 when $\tau \searrow t$.

It is preferable to write (6-8) as a differential equation for f_t rather than f_t^{-1}. Since f_t^{-1} has a nonzero derivative with respect to w, it follows by the implicit function theorem that f_t is differentiable, and

$$\frac{\partial f_t^{-1}(w)}{\partial t} + \frac{\partial f_t^{-1}(w)}{\partial w} \frac{\partial f_t(z)}{\partial t} = 0$$

when $w = f_t(z)$. Substitution from (6-8) yields

$$\frac{\partial f_t(z)}{\partial t} = -f_t'(z)z \frac{\lambda(t) + z}{\lambda(t) - z}. \tag{6-9}$$

This is Löwner's famous differential equation.

6-3 PROOF OF $|a_3| \leq 3$

We shall write

$$f_t(z) = e^{-t}[z + a_2(t)z^2 + a_3(t)z^3 + \cdots].$$

To see that this series can be differentiated termwise it is sufficient to express f_t and its derivatives $D^n f_t$ with respect to z as Cauchy integrals, for instance, over $|z| = \frac{1}{2}$. The integrals can be differentiated with respect to t under the integral sign, and we may conclude that $\partial D^n f_t / \partial t = D^n(\partial f_t / \partial t)$. Consequently, $a_n'(t)$ exists, and we obtain in shorter notation

$$\frac{\partial f_t}{\partial t} = -e^{-t}(z + a_2 z^2 + a_3 z^3 + \cdots) + e^{-t}(a_2' z^2 + a_3' z^3 + \cdots).$$

On the right-hand side of (6-9) we substitute

$$f_t'(z) = e^{-t}(1 + 2a_2 z + 3a_3 z^2 + \cdots),$$

and

$$\frac{\lambda + z}{\lambda - z} = 1 + \frac{2z}{\lambda} + \frac{2z^2}{\lambda^2} + \cdots.$$

Comparison of the coefficients leads to

$$a_2' - a_2 = -2a_2 - 2\lambda^{-1}$$
$$a_3' - a_3 = -3a_3 - 4a_2\lambda^{-1} - 2\lambda^{-2},$$

which we can rewrite as

$$\frac{d}{dt}(a_2 e^t) = -2\lambda^{-1}e^t$$

$$\frac{d}{dt}(a_3 e^{2t}) = (-4a_2\lambda^{-1} - 2\lambda^{-2})e^{2t}.$$

Recall that $a_2 = a_3 = 0$ for $t = 0$. Therefore, integration yields

$$a_2(\tau)e^\tau = -2\int_0^\tau e^t\lambda^{-1}\,dt$$
$$a_3(\tau)e^{2\tau} = 4\left(\int_0^\tau e^t\lambda^{-1}\,dt\right)^2 - 2\int_0^\tau e^{2t}\lambda^{-2}\,dt.$$

We set $\lambda = e^{i\theta}$ and take real parts in the second equation. This gives

$$\mathrm{Re}\,a_3(\tau)e^{2\tau} = 4\left(\int_0^\tau e^t\cos\theta\,dt\right)^2 - 4\left(\int_0^\tau e^t\sin\theta\,dt\right)^2$$
$$- 2\int_0^\tau e^{2t}(2\cos^2\theta - 1)\,dt.$$

The Schwarz inequality leads to the estimate

$$\left(\int_0^\tau e^t\cos\theta\,dt\right)^2 < e^\tau\int_0^\tau e^t\cos^2\theta\,dt,$$

and we obtain

$$\mathrm{Re}\,a_3(\tau)e^{2\tau} < 4\int_0^\tau e^t(e^\tau - e^t)\cos^2\theta\,dt + e^{2\tau} - 1.$$

Since $t < \tau$ and $\cos^2\theta \le 1$, it follows that

$$\mathrm{Re}\,a_3(\tau)e^{2\tau} < 4\int_0^\tau e^t(e^\tau - e^t)\,dt + e^{2\tau} - 1 = 3e^{2\tau} - 4e^\tau + 1 < 3e^{2\tau}.$$

We have proved that $\mathrm{Re}\,a_3(\tau) < 3$ for all τ. Hence $\mathrm{Re}\,a_3 < 3$ for the original function $f(z) = e^{-t_0}(z + a_2 z^2 + a_3 z^3 + \cdots)$. On applying the result to $e^{-i\alpha}f(e^{i\alpha}z)$ we conclude that $|a_3| < 3$ when f is analytic on the boundary, and hence $|a_3| \le 3$ for an arbitrary normalized univalent function.

NOTES   The basic reference is Löwner [36]. The original proof uses Löwner's lemma (Sec. 1-4). The method has been used extensively for more general problems and in combination with other variational methods.

7

THE SCHIFFER VARIATION

7-1 VARIATION OF THE GREEN'S FUNCTION

Let Ω be a region in the complex plane with Green's function $g(z,\zeta)$. We wish to find out what happens to g when Ω is replaced by a nearby region Ω^*. An obvious way would be to express g as an integral over the boundary. This has the serious drawback that it requires the boundary to be smooth. If the variation is to be used for the solution of an extremal problem, it must be applicable in a situation in which the boundary is not known to be smooth. It was to overcome this difficulty that Schiffer [57] devised his method of *interior variation*. This method is very simple in principle, but the computations require some patience.

Consider a point $z_0 \in \Omega$ and a circle c with small radius centered at z_0. The point z_0 will be kept fixed while ρ tends to zero. We also fix a real number α. The function

$$z^* = z + \rho^2 e^{i\alpha}(z - z_0)^{-1} \tag{7-1}$$

maps the outside of c on the complement of a line segment of length 4ρ with midpoint z_0 and inclination $\alpha/2$. The complement E of Ω is mapped on a set E^* whose complement is in turn a region Ω^*. We denote the Green's function of Ω^* by g^*. Our aim is to find an asymptotic expression for $\delta g(z,\zeta) = g^*(z,\zeta) - g(z,\zeta)$ when ρ approaches zero.

We assume the existence of $g(z,\zeta)$. The existence of $g^*(z,\zeta)$ is implicit in the proof of our first lemma, which will serve to give some crude preliminary estimates.

Lemma 7-1 The functions $g^*(z,\zeta)$ and their partial derivatives are uniformly bounded when z and ζ range over compact sets in Ω and $|z - \zeta|$ is bounded away from zero.

PROOF Because $g(z,\zeta)$ is symmetric, it is sufficient to prove the lemma for a fixed ζ. The inversion $z' = 1/(z - \zeta)$ maps Ω on a region Ω' in the extended plane. Clearly, $g'(z') = g(\zeta + 1/z', \zeta)$ is the Green's function of Ω' with pole at ∞. The notations $\Omega^{*\prime}$ and $g^{*\prime}$ are self-explanatory. It will be sufficient to prove that $g^{*\prime}$ and its derivatives are uniformly bounded on every compact set in Ω'. Observe that such a compact set is contained in $\Omega^{*\prime}$ for all sufficiently small ρ, and the uniform bound is to be valid for $\rho \leq \rho_0$, for example.

Recall that $g'(z') = \log |z'| + \gamma' + O(1)$ as $z' \to \infty$. The Robin constant γ' is connected with the transfinite diameter of the complement E' of Ω' by $\gamma' = -\log d_\infty$. By definition d_∞ is the limit of d_n, and d_n is the maximum geometric mean of the mutual distances $|z_i' - z_j'|$ of n points on E'.

With obvious notations we obtain from (7-1)

$$z_i^{*\prime} - z_j^{*\prime} = (z_i' - z_j') \frac{z_i^{*\prime} z_j^{*\prime}}{z_i' z_j'} \left[1 - \frac{\rho^2 e^{i\alpha}}{(z_i - z_0)(z_j - z_0)} \right]. \qquad (7\text{-}2)$$

An easy calculation shows that $|z_i^{*\prime}|/|z_i'| = 1 + O(\rho^2)$, uniformly for $z_i' \in E'$, and the same estimate applies to the last factor in (7-2). It readily follows that $d_\infty^* = d_\infty[1 + O(\rho^2)]$ and $\gamma^{*\prime} = \gamma' + O(\rho^2)$.

Ω' can be exhausted by regions Ω_n' with smooth boundaries. To save notation we assume temporarily that Ω' itself has a smooth boundary. Routine use of Green's formula gives

$$g'(z) = \gamma' - \frac{1}{2\pi} \int_{\partial\Omega'} \frac{\partial g'}{\partial n_t} \log |t - z| \, |dt|. \qquad (7\text{-}3)$$

Here n_t is the outer normal, so that $\partial g'/\partial n_t < 0$. We conclude from (7-3) that

$$g'(z) \leq \gamma' + \max_{t \in \partial\Omega'} \log |t - z|.$$

Because this holds for all Ω_n', it also holds for an Ω' with arbitrary boundary. When applied to $\Omega^{*\prime}$ the inequality shows that the $g^{*\prime}$ are uniformly bounded on every compact set. Standard use of the Poisson integral shows that the same is true of the derivatives, and the lemma is proved.

We shall write the difference $g^*(z,\zeta) - g(z,\zeta)$ as the sum of $\delta_1(z,\zeta) = g^*(z^*,\zeta^*) - g(z,\zeta)$ and $\delta_2(z,\zeta) = g^*(z,\zeta) - g^*(z^*,\zeta^*)$. To begin with, z and ζ shall lie outside the circle c, and z^*, ζ^* are given by (7-1). As a function of z the difference $\delta_1(z,\zeta)$ is defined and harmonic in the part of Ω outside c. It vanishes on $\partial\Omega$ and has no singularity at ζ. Therefore, Green's formula yields

$$\delta_1(z,\zeta) = -\frac{1}{2\pi}\int_c\left[\delta_1(t,\zeta)\frac{\partial g(t,z)}{\partial n_t} - g(t,z)\frac{\partial\delta_1(t,\zeta)}{\partial n_t}\right]|dt|. \tag{7-4}$$

Because

$$\int_c\left[g(t,\zeta)\frac{\partial g(t,z)}{\partial n_t} - g(t,z)\frac{\partial g(t,\zeta)}{\partial n_t}\right]|dt| = 0,$$

we can rewrite (7-4) as

$$\delta_1(z,\zeta) = -\frac{1}{2\pi}\int_c\left[g^*(t^*,\zeta^*)\frac{\partial g(t,z)}{\partial n_t} - g(t,z)\frac{\partial g^*(t^*,\zeta^*)}{\partial n_t}\right]|dt|. \tag{7-5}$$

As before, the formula is first proved for an Ω_n with a smooth boundary, but it remains true for an arbitrary Ω.

For convenience we introduce the notations

$$\Gamma(z,\zeta) = \frac{\partial g(z,\zeta)}{\partial z} = \frac{1}{2}\left(\frac{\partial g}{\partial x} - i\frac{\partial g}{\partial y}\right)$$

$$\Gamma^*(z,\zeta) = \frac{\partial g^*(z,\zeta)}{\partial z}.$$

These functions are analytic in z, and one verifies that formula (7-5) takes the form

$$\delta_1(z,\zeta) = \frac{1}{\pi}\operatorname{Im}\int_c\left[g^*(t^*,\zeta^*)\Gamma(t,z) - g(t,z)\Gamma^*(t^*,\zeta^*)\frac{dt^*}{dt}\right]dt, \tag{7-6}$$

where the integral is in the positive sense of the circle.

In the first term on the right in (7-6) we insert the development

$$g^*(t^*,\zeta^*) = g^*(z_0,\zeta^*) + \Gamma^*(z_0,\zeta^*)(t^* - z_0) + \bar{\Gamma}^*(z_0,\zeta^*)(\bar{t}^* - \bar{z}_0) + O(\rho^2).$$

Here the remainder involves second-order derivatives of $g^*(t,\zeta^*)$ for $|t - z_0| \leq 2\rho$. By virtue of Lemma 7-1 the estimate is uniformly valid when ζ stays away from z_0. The integral can now be evaluated by residues.

$\Gamma(t,z)$ is regular for t inside c; $t^* - z_0$ has a pole at z_0 with residue $\rho^2 e^{i\alpha}$; and on c we have $\bar{t}^* - \bar{z}_0 = \rho^2/(t - z_0) + e^{-i\alpha}(t - z_0)$, which has the residue ρ^2. We obtain

$$\int_c g^*(t^*,\zeta^*)\Gamma(t,z)\,dt = 2\pi i\Gamma(z_0,z)[\Gamma^*(z_0,\zeta^*)e^{i\alpha} + \bar{\Gamma}^*(z_0,\zeta^*)]\rho^2 + O(\rho^3).$$
(7-7)

In the second term on the right-hand side of (7-6) all three factors have to be expanded. The expansions are, for $|t - z_0| = \rho$,

$$g(t,z) = g(z_0,z) + \Gamma(z_0,z)(t - z_0) + \bar{\Gamma}(z_0,z)\frac{\rho^2 e^{i\alpha}}{t - z_0} + O(\rho^2)$$

$$\Gamma^*(t^*,\zeta^*) = \Gamma^*(z_0,\zeta^*) + \frac{\partial\Gamma^*(z_0,\zeta^*)}{\partial z}\left(t - z_0 + \frac{\rho^2 e^{i\alpha}}{t - z_0}\right) + O(\rho^2)$$

$$\frac{dt^*}{dt} = 1 - \frac{\rho^2 e^{i\alpha}}{(t - z_0)^2}.$$

The product of the principal parts has the residue

$$\rho^2[\bar{\Gamma}(z_0,z) - e^{i\alpha}\Gamma(z_0,z)]\Gamma^*(z_0,\zeta^*),$$

and we find

$$\int_c g(t,z)\Gamma^*(t^*,\zeta^*)\,dt^* = 2\pi i\rho^2[\bar{\Gamma}(z_0,z) - e^{i\alpha}\Gamma(z_0,z)]\Gamma^*(z_0,\zeta^*) + O(\rho^3). \quad (7\text{-}8)$$

Substitution of (7-7) and (7-8) in (7-6) gives

$$\delta_1(z,\zeta) = 4\rho^2\,\text{Re}\,[\Gamma(z_0,z)\Gamma^*(z_0,\zeta^*)e^{i\alpha}] + O(\rho^3). \quad (7\text{-}9)$$

The expansion is valid for arbitrary regions, and the estimate of the remainder is uniform as long as z and ζ stay in compact sets that do not include z_0.

Next we write down the development

$$\begin{aligned}\delta_2(z,\zeta) &= g^*(z,\zeta) - g^*(z^*,\zeta^*)\\ &= -2\,\text{Re}\,[\Gamma^*(z,\zeta)(z^* - z) + \Gamma^*(\zeta,z)(\zeta^* - \zeta)] + O(\rho^4)\\ &= -2\rho^2\,\text{Re}\left\{e^{i\alpha}\left[\frac{\Gamma^*(z,\zeta)}{z - z_0} + \frac{\Gamma^*(\zeta,z)}{\zeta - z_0}\right]\right\} + O(\rho^4), \quad (7\text{-}10)\end{aligned}$$

which is uniformly valid provided that z and ζ stay away from each other and from z_0.

As a crude estimate (7-9) and (7-10) yield $g^*(z,\zeta) - g(z,\zeta) = O(\rho^2)$, and by differentiation $\Gamma^*(z,\zeta) - \Gamma(z,\zeta) = O(\rho^2)$. The proof requires z to stay away from z_0, but we can use the maximum principle to conclude that the estimate continues to hold near z_0. It is also obvious from Lemma 7-1 that $\Gamma^*(z,\zeta^*) - \Gamma^*(z,\zeta) = O(\rho^2)$. As a result of these estimates the asterisks in (7-9) and (7-10) can be dropped, and we have proved a variational formula for Green's function.

Lemma 7-2 There exists a two-parameter family of regions Ω^* such that

$$g^*(z,\zeta) - g(z,\zeta) = 2\rho^2 \operatorname{Re} \left\{ e^{i\alpha} \left[2\Gamma(z_0,z)\Gamma(z_0,\zeta) \right. \right.$$

$$\left. \left. - \frac{\Gamma(z,\zeta)}{z - z_0} - \frac{\Gamma(\zeta,z)}{\zeta - z_0} \right] \right\} + O(\rho^3). \quad (7\text{-}11)$$

More precisely, to every compact set $K \subset \Omega$ there exist ρ_0 and M such that the left-hand member of (7-11) is defined and the remainder is $< M\rho^3$ for all $\rho < \rho_0$ and all $z,\zeta,z_0 \in K$.

The statement has not yet been proved when z, ζ, and z_0 are close to each other, but it easily follows by use of the maximum principle. In fact, neither the left-hand member nor the expression in braces has any singularity. To check that this is so, let us write

$$g(z,\zeta) = -\log |z - \zeta| + \gamma(z,\zeta)$$
$$\Gamma(z,\zeta) = -\tfrac{1}{2}(z - \zeta)^{-1} + \gamma_1(z,\zeta).$$

One finds that the expression inside the square brackets in (7-11) can be written as

$$- \frac{\gamma_1(z,\zeta) - \gamma_1(z_0,\zeta)}{z - z_0} - \frac{\gamma_1(\zeta,z) - \gamma_1(z_0,z)}{\zeta - z_0} + 2\gamma_1(z_0,z)\gamma_1(z_0,\zeta).$$

It is obviously regular for all values of the variables.

7-2 VARIATION OF THE MAPPING FUNCTION

We now make the additional assumption that Ω is simply connected and that $0 \in \Omega$. By the Riemann mapping theorem there exists a unique conformal mapping φ of Ω onto the unit disk such that $\varphi(0) = 0$ and $\varphi'(0) > 0$. The mapping function of Ω^* is denoted by φ^*.

We are going to apply (7-11) with $z = 0$, $z_0 = \zeta_0$. In other words, we start from the formula

$$g^*(0,\zeta) - g(0,\zeta) = 2\rho^2 \operatorname{Re} \left\{ e^{i\alpha} \left[2\Gamma(\zeta_0,0)\Gamma(\zeta_0,\zeta) \right. \right.$$

$$\left. \left. + \frac{1}{\zeta_0} \Gamma(0,\zeta) - \frac{1}{\zeta - \zeta_0} \Gamma(\zeta,0) \right] \right\} + O(\rho^3). \quad (7\text{-}12)$$

Green's function with pole at the origin is $g(\zeta,0) = -\log|\varphi(\zeta)|$. The general value is

$$g(\zeta_0,\zeta) = -\log\left|\frac{\varphi(\zeta_0) - \varphi(\zeta)}{1 - \overline{\varphi(\zeta)}\varphi(\zeta_0)}\right|,$$

and differentiation yields

$$\Gamma(\zeta_0,\zeta) = -\frac{1}{2}\left[\frac{\varphi'(\zeta_0)}{\varphi(\zeta_0) - \varphi(\zeta)} + \frac{\overline{\varphi(\zeta)}\varphi'(\zeta_0)}{1 - \overline{\varphi(\zeta)}\varphi(\zeta_0)}\right].$$

We need the special values

$$\Gamma(\zeta,0) = -\frac{1}{2}\frac{\varphi'(\zeta)}{\varphi(\zeta)}, \qquad \Gamma(0,\zeta) = -\frac{1}{2}\left[-\frac{\varphi'(0)}{\varphi(\zeta)} + \overline{\varphi(\zeta)}\varphi'(0)\right].$$

When these expressions are substituted in (7-12), we obtain

$$\log\frac{|\varphi^*(\zeta)|}{|\varphi(\zeta)|} = -\rho^2\,\mathrm{Re}\,[e^{i\alpha}A(\zeta) + e^{-i\alpha}B(\zeta)] + O(\rho^3), \qquad (7\text{-}13)$$

where
$$A(\zeta) = \frac{\varphi'(\zeta_0)^2}{\varphi(\zeta_0)[\varphi(\zeta_0) - \varphi(\zeta)]} + \frac{\varphi'(0)}{\zeta_0\varphi(\zeta)} + \frac{\varphi'(\zeta)}{(\zeta - \zeta_0)\varphi(\zeta)}$$

$$B(\zeta) = \frac{\varphi(\zeta)\overline{\varphi'(\zeta_0)}^2}{\varphi(\zeta_0)[1 - \varphi(\zeta)\varphi(\zeta_0)]} - \frac{\varphi(\zeta)\varphi'(0)}{\zeta_0}. \qquad (7\text{-}14)$$

We have been careful to distribute the terms so that $A(\zeta)$ and $B(\zeta)$ are analytic in ζ.

We shall now add the conjugate harmonic functions on both sides of (7-13). If we want the same remainder estimate to hold for the imaginary part, we must choose the additive constant so that the imaginary part vanishes at the origin. We find in this way

$$\log\frac{\varphi^*(\zeta)}{\varphi(\zeta)} = -\rho^2\left\{e^{i\alpha}\left[A(\zeta) - \frac{A(0)}{2}\right] + e^{-i\alpha}\left[B(\zeta) + \frac{A(0)}{2}\right]\right\} + O(\rho^3),$$

and after exponentiation

$$\varphi^*(\zeta) = \varphi(\zeta)\left(1 - \rho^2\left\{e^{i\alpha}\left[A(\zeta) - \frac{A(0)}{2}\right]\right.\right.$$
$$\left.\left. + e^{-i\alpha}\left[B(\zeta) + \frac{\overline{A(0)}}{2}\right]\right\} + O(\rho^3)\right). \qquad (7\text{-}15)$$

We observe that $A(0)$ must be computed as a limit and has the value

$$A(0) = \frac{\varphi'(\zeta_0)^2}{\varphi(\zeta_0)^2} - \frac{\varphi''(0)}{\zeta_0\varphi'(0)} - \frac{1}{\zeta_0^2}.$$

For $\zeta_0 = 0$ still another limit must be formed. The estimate in (7-15) is uniform as long as ζ and ζ_0 stay in compact sets.

In most applications it is more convenient to deal with the inverse function f of φ and compare it with the inverse f^* of φ^*. In fact, the primary problem is to study the variation of a *schlicht* function in the unit disk. For this purpose we substitute $\zeta = f(z)$ and $\zeta_0 = f(z_0)$ in (7-14) and (7-15). To simplify the notation we write $A(z)$ and $B(z)$ instead of $A[f(z)]$ and $B[f(z)]$. Thus

$$A(z) = \frac{1}{z_0(z_0 - z)f'(z_0)^2} + \frac{1}{f'(0)zf(z_0)} + \frac{1}{zf'(z)[f(z) - f(z_0)]}$$

$$B(z) = \frac{z}{\bar{z}_0(1 - z\bar{z}_0)\overline{f'(z_0)}^2} - \frac{z}{f'(0)\overline{f(z_0)}}, \tag{7-16}$$

and (7-15) becomes

$$\varphi^*[f(z)] = z \left(1 - \rho^2 \left\{ e^{i\alpha} \left[A(z) - \frac{A(0)}{2} \right] \right. \right.$$
$$\left. \left. + e^{-i\alpha} \left[B(z) + \frac{\overline{A(0)}}{2} \right] \right\} + O(\rho^3) \right). \tag{7-17}$$

When both sides of (7-17) are taken as argument in f^*, we obtain

$$f(z) = f^*(z) - zf^{*\prime}(z)\rho^2 \left\{ e^{i\alpha} \left[A - \frac{A(0)}{2} \right] + e^{-i\alpha} \left[B + \frac{\overline{A(0)}}{2} \right] \right\} + O(\rho^3). \tag{7-18}$$

The estimate remains uniform, for one shows with the aid of Lemma 7-1 that the derivatives of f^* are bounded on any compact set. Hence (7-18) yields the crude estimates $f^* - f = O(\rho^2)$ and $f^{*\prime} - f' = O(\rho^2)$. In (7-18) we can therefore replace $f^{*\prime}$ by f', so that the variational formula becomes

$$f^*(z) - f(z) = zf'(z)\rho^2 \left\{ e^{i\alpha} \left[A - \frac{A(0)}{2} \right] + e^{-i\alpha} \left[B + \frac{\overline{A(0)}}{2} \right] \right\} + O(\rho^3).$$

If we wish the mapping function to be normalized, $f'(0) = 1$, we must divide f^* by

$$f^{*\prime}(0) = 1 + \rho^2 \operatorname{Re}[A(0)e^{i\alpha}] + O(\rho^3).$$

For simplicity the normalized function will again be denoted by f^*, and we obtain for the normalized variation

$$f^*(z) - f(z) = \rho^2 zf'(z)\{A(z)e^{i\alpha} + B(z)e^{-i\alpha} - i\operatorname{Im}[A(0)e^{i\alpha}]\}$$
$$- \rho^2 f(z)\operatorname{Re}[A(0)e^{i\alpha}] + O(\rho^3). \tag{7-19}$$

Here $A(z)$ and $B(z)$ have to be substituted from (7-16) with the slight simplification $f'(0) = 1$, and

$$A(0) = \frac{1}{z_0^2 f'(z_0)^2} + \frac{f''(0)}{f(z_0)} - \frac{1}{f(z_0)^2}.$$

7-3 THE FINAL THEOREM

There are still other variations of a more elementary nature, and they can be added to Schiffer's variation in an effort to gain generality or simplify the result. First, we may replace $f(z)$ by $e^{-i\gamma}f(e^{i\gamma}z)$, where γ is a small real number. One finds quite easily that the resulting variation is

$$\delta f = i\gamma[zf'(z) - f(z)] + O(\gamma^2). \tag{7-20}$$

We shall choose $\gamma = \rho^2 \operatorname{Im}[A(0)e^{i\alpha}]$ and add the new variation to (7-19). In this way we obtain

$$f^*(z) - f(z) = \rho^2\{[A(z)e^{i\alpha} + B(z)e^{-i\alpha}]zf'(z) - A(0)e^{i\alpha}f(z)\} + O(\rho^3). \tag{7-21}$$

This variation is the one given by Schiffer [15] [compare with his formula (A3.30) which is not yet normalized].

There is a second elementary variation due to Marty [37]. For small complex ϵ consider

$$f^*(z) = f'(\epsilon)^{-1}\left[f\left(\frac{z + \epsilon}{1 + \bar{\epsilon}z}\right) - f(\epsilon) \right](1 - |\epsilon|^2)^{-1},$$

which is normalized and univalent. The variation is easily seen to be of the form

$$\delta f = \epsilon[f'(z) - 1 - f''(0)f(z)] - \bar{\epsilon}z^2 f'(z) + O(|\epsilon|^2). \tag{7-22}$$

We choose $\epsilon = -\rho^2 e^{i\alpha}f(z_0)^{-1}$ and add to (7-21). The terms with $f''(0)$ cancel, and there are other simplifications as well. The resulting formula is particularly neat, and we shall formalize it as a theorem.

Theorem 7-1 There exists a normalized variation of the form

$$\delta f(z) = \rho^2[L(z)e^{i\alpha} + M(z)e^{-i\alpha}] + O(\rho^3) \tag{7-23}$$

with

$$L(z) = \frac{zf'(z)}{z_0(z_0 - z)f'(z_0)^2} - \frac{f(z)}{z_0^2 f'(z_0)^2} + \frac{f(z)^2}{f(z_0)^2[f(z) - f(z_0)]}$$

$$M(z) = \frac{z^2 f'(z)}{\bar{z}_0(1 - z\bar{z}_0)\overline{f'(z_0)}^2}. \tag{7-24}$$

The estimate is uniform as long as z and z_0 stay in compact sets and z_0 stays away from zero.

Although we give preference to (7-23), it should be remembered that other combinations of (7-19), (7-20), and (7-22) may also be used.

7-4 THE SLIT VARIATION

Let

$$K(z) = \frac{z}{(1 + z)^2}$$

be the Koebe function which maps $|z| < 1$ on the plane slit along the real axis from $\frac{1}{4}$ to $+\infty$. Because the image region is star-shaped with respect to the origin, we can form

$$E_t(z) = K^{-1}[e^{-t}K(z)]$$

for all $t > 0$. The asymptotic behavior of E_t for small t is found by the following calculation:

$$E_t(z) = K^{-1}[(1 - t)K(z)] + O(t^2) = z - \frac{K(z)}{K'(z)} t + O(t^2)$$

$$= z - \frac{z(1 + z)}{1 - z} t + O(t^2).$$

Starting from a normalized *schlicht* f we form $f^*(z) = f[e^{-i\gamma}E_t(e^{i\gamma}z)]$, which is again *schlicht*. It has the development

$$f^*(z) = f(z) - zf'(z) \frac{1 + e^{i\gamma}z}{1 - e^{i\gamma}z} t + O(t^2).$$

To normalize we have to divide by $f^{*\prime}(0) = 1 - t + O(t^2)$. We conclude that there exists a normalized variation of the form

$$\delta f = \left[f(z) - zf'(z) \frac{1 + e^{i\gamma}z}{1 - e^{i\gamma}z} \right] t + O(t^2). \tag{7-25}$$

It differs fundamentally from the earlier variations by the fact that t cannot be replaced by $-t$. For this reason, when applied to an extremal problem, it gives rise to an inequality rather than an equation.

NOTES Schiffer's idea of using interior variations was a breakthrough in the theory of univalent functions. It has permeated much of the postwar literature in this field. Schiffer's own retrospective account is in his appendix to Courant [15].

PROPERTIES OF THE EXTREMAL FUNCTIONS

8-1 THE DIFFERENTIAL EQUATION

We reintroduce the notation $f(z) = \sum_{1}^{\infty} a_n z^n$, $a_1 = 1$, for a normalized *schlicht* function and address ourselves to the problem of maximizing $|a_n|$. If $f(z)$ is replaced by $e^{-i\gamma}f(e^{i\gamma}z)$, the coefficients become $a_n e^{(n-1)i\gamma}$. For this reason it is an equivalent problem to maximize Re a_n, and the maximum of Re a_n will occur when a_n is positive. The existence of an extremal function is trivial, and we shall use the results of the preceding chapter to derive some of its properties.

We return to the notations of Theorem 7-1 and write

$$L(z) = \sum_{1}^{\infty} L_n z^n$$

$$M(z) = \sum_{1}^{\infty} M_n z^n.$$

It is clear from this theorem that the function which maximizes Re a_n

must satisfy

$$\text{Re}\,(L_n e^{i\alpha} + M_n e^{-i\alpha}) = 0$$

for all real α, and this is so if and only if

$$L_n + \bar{M}_n = 0. \tag{8-1}$$

In order to analyze this condition we must make it more explicit. It is only the last term in the expression for $L(z)$ whose development causes some difficulty. For the purpose of obtaining at least semiexplicit results we shall write

$$\frac{tf(z)^2}{1 - tf(z)} = \sum_{2}^{\infty} S_n(t)z^n,$$

where the S_n are certain polynomials in t of degree $n - 1$ with leading coefficient 1 and zero constant term. With this notation (7-24) yields

$$L_n = \frac{1}{z_0{}^2 f'(z_0)^2}\left[(n - 1)a_n + \sum_{k=1}^{n-1} \frac{ka_k}{z_0^{n-k}} \right] - \frac{1}{f(z)^2} S_n\left(\frac{1}{f(z_0)} \right)$$

$$\bar{M}_n = \frac{1}{z_0{}^2 f'(z_0)^2}\left[\sum_{k=1}^{n-1} k\bar{a}_k z_0^{n-k} \right].$$

To conform with standard usage we shall write

$$Q_n(z) = \sum_{k=1}^{n-1} \frac{ka_k}{z^{n-k}} + (n - 1)a_n + \sum_{k=1}^{n-1} k\bar{a}_k z^{n-k}$$

$$P_n(w) = S_n\left(\frac{1}{w} \right). \tag{8-2}$$

If we replace z_0 by z, condition (8-1) takes the form

$$\frac{P_n[f(z)]f'(z)^2}{f(z)^2} = \frac{Q_n(z)}{z^2}.$$

In other words, we have shown that the extremal function $w = f(z)$ is a solution of the differential equation

$$\frac{P_n(w)w'^2}{w^2} = \frac{Q_n(z)}{z^2}. \tag{8-3}$$

The consequences of this remarkable relation will be studied later. For the moment we recall that $a_n > 0$, as already mentioned. In particular, a_n is real, so that the symmetry of the coefficients of the rational function $Q_n(z)$ makes it real on $|z| = 1$. Hence the zeros of Q_n lie on the unit circle or are pairwise symmetric to it. The only poles are at 0 and ∞, and they are of order $n - 1$. As for $P_n(w)$ the most important feature is the pole of order $n - 1$ at the origin.

We have chosen to maximize Re a_n only as a typical example. It is in no way more difficult to maximize an arbitrary real-valued differentiable function $F(a_2, \ldots, a_n)$, and we prefer to consider the problem in this generality. We denote its complex derivatives by $\partial F/\partial a_k = F_k$, and in order to exclude extraneous solutions we assume that they are never simultaneously zero. The variational condition is obviously

$$\sum_{k=2}^{n} (F_k L_k + \bar{F}_k \bar{M}_k) = 0,$$

and this leads to a differential equation

$$\frac{P(w)w'^2}{w^2} = \frac{Q(z)}{z^2}, \tag{8-4}$$

where P and Q are rational functions similar to P_n and Q_n.

One sees at once that P and Q are of the form

$$P(w) = \sum_{1}^{n-1} A_k w^{-k}$$

$$Q(z) = \sum_{-(n-1)}^{n-1} B_k z^k, \tag{8-5}$$

where $B_0 = \sum_{2}^{n} (k - 1)F_k a_k$, $B_{-k} = \bar{B}_k$ for $k \neq 0$, and

$$B_k = \sum_{h=1}^{n-k} h F_{h+k} a_h \tag{8-6}$$

for $k > 0$. Therefore, Q is real on $|z| = 1$ if B_0 is real. That this is so is a consequence of the variation (7-20). According to this formula there is a permissible variation $\delta f = i\gamma[zf'(z) - f(z)]$ for real γ. The corresponding variation of the coefficients is $\delta a_k = i\gamma(k - 1)a_k$, and if $F(a_2, \ldots, a_n)$ is a maximum, we must have Re $\Sigma F_k \delta a_k = 0$, and hence Im $B_0 = 0$.

Because the differential equation (8-4) has a solution $w = f(z)$ with $f'(0) = 1$, the highest coefficients must be equal, $A_{n-1} = B_{n-1}$. In the case of equation (8-3) these coefficients are $\neq 0$. In the general case we do not know this, but under the assumption that not all F_k are zero Q cannot be identically zero and therefore has a highest nonvanishing coefficient. With a change of notation we may therefore assume that $B_{n-1} \neq 0$.

It will be important for our later discussion to know that Q is not only real on the unit circle, but actually ≥ 0. This can be seen by use of the slit variation (7-25). When expanded this variation can be written

$$\delta f = \left[\sum_1^\infty a_n z^n - \left(\sum_1^\infty k a_k z^k \right) \left(1 + 2 \sum_1^\infty e^{hi\gamma} z^h \right) \right] t + O(t^2),$$

and we find $\delta a_n = -e_n t + O(t^2)$ with

$$e_n = (n-1)a_n + 2 \sum_{k=1}^{n-1} k a_k e^{(n-k)i\gamma}.$$

Recall that this is a valid variation only when $t > 0$. If $F(a_2, \ldots, a_n)$ is a maximum, Re $\Sigma F_k \delta a_k$ must be ≤ 0 for $t > 0$, and this implies Re $\Sigma F_k e_k \geq 0$. On the other hand, with the aid of (8-2), (8-5), and (8-6), one finds Re $\Sigma F_k e_k = Q(e^{-i\gamma})$. Since γ is an arbitrary real number, it follows that $Q(z) = 0$ for $|z| = 1$.

8-2 TRAJECTORIES

Equation (8-4) can be written

$$\frac{P(w)\, dw^2}{w^2} = \frac{Q(z)\, dz^2}{z^2}, \tag{8-7}$$

and both sides may be regarded as *quadratic differentials*. The fact that $f(z)$ satisfies the differential equation means that (8-7) becomes an identity when we substitute $w = f(z)$, $dw = f'(z)\, dz$. It is clear that any zero of $Q(z)$ inside the unit circle is mapped on a zero of $P(w)$. Another consequence is that any arc on which $Q\, dz^2/z^2$ has a constant argument is mapped on an arc on which $P\, dw^2/w^2$ has the same constant argument. Such arcs are called *trajectories*. In particular, we speak of a horizontal trajectory if $Q\, dz^2/z^2 \geq 0$, and a vertical trajectory if $Q\, dz^2/z^2 \leq 0$. For instance, because $Q \geq 0$ on $|z| = 1$, the unit circle is a vertical trajectory, but it is premature to speak of its image under f. Nevertheless, the study of the trajectories, and especially the vertical trajectories, should give us valuable information about the mapping.

For somewhat greater generality we shall denote the quadratic differential under consideration by $\varphi(w)\, dw^2$. Here $\varphi(w)$ will be rational, and the zeros and poles of φ are referred to as the singularities of the quadratic differential. In order to stress the geometric point of view we introduce a metric $ds^2 = |\varphi(w)||dw|^2$, which is euclidean except at the singularities. The trajectories are the geodesics of this metric.

In order to investigate the trajectories near a given point w_0 we introduce an auxiliary variable

$$\zeta = \int_{w_0}^{w} \sqrt{\varphi(w)}\, dw. \tag{8-8}$$

The vertical trajectories correspond to the lines Re $\zeta(w)$ = constant, but the situation is complicated by the fact that ζ is not single-valued.

We denote by m the order of φ at w_0, $m > 0$ for a zero, $m < 0$ for a pole. We have to distinguish several cases.

1) $m = 0$. We can choose a single-valued branch of $\sqrt{\varphi(w)}$ in a neighborhood of w_0. The choice of branch does not matter since a change of sign would merely replace ζ by $-\zeta$. The function $\zeta(w)$ has a simple zero at w_0, and therefore there is a single line Re $\zeta = 0$ passing through w_0. If we prefer, we can say that there are two vertical trajectories issuing from w_0 in opposite directions.

2) $m > 0$. We can perform the integration in (8-8) to obtain

$$\zeta = (w - w_0)^{m/2+1}\psi(w), \tag{8-9}$$

where ψ is analytic and $\neq 0$ near w_0. The sign is ambiguous, but of no importance. The directions of the lines Re $\zeta = 0$ are given by

$$\left(\frac{m}{2} + 1\right) \arg w + \arg \psi(w_0) = \frac{\pi}{2} + n\pi,$$

with integral n. There are thus $m + 2$ equally spaced vertical trajectories issuing from w_0.

3) $m = -1$. This is similar to the preceding case, for it is still true that $\zeta = 0$ for $w = w_0$. There is a vertical trajectory from w_0 in one direction only.

4) $m = -(2k + 1)$, $k > 0$. The lower bound must be replaced by some other constant. Integration still leads to a development similar to (8-9), but now $\zeta = \infty$ for $w = w_0$. Consequently, all lines Re $\zeta = c$ pass through w_0. There are infinitely many vertical trajectories from w_0, and they are tangent to $2k - 1$ equally spaced directions.

5) $m = -2$. This case will never occur in our applications, and we leave the discussion to the reader.

6) $m = -2k$, $k > 1$. There is a complication due to the fact that the development of ζ may contain a logarithmic term. In place of (8-9) we obtain

$$\zeta = (w - w_0)^{1-k}\psi(w) + \alpha \log (w - w_0),$$

with $\psi(w_0) \neq 0$ and constant α. Let us write $w - w_0 = re^{i\theta}$, $\psi(w_0) = \rho e^{i\theta_0}$, $\alpha = \alpha_1 + i\alpha_2$. The lines Re ζ = constant are loci of the form

$$r^{1-k}\{\cos [(k - 1)(\theta - \theta_0)]\} + r\psi(r,\theta) + \alpha_1 \log r - \alpha_2\theta = c,$$

where $\psi(r,\theta)$ is differentiable. If the equation is multiplied by r^{k-1}, it will be satisfied for $r = 0$ and $2k - 2$ equally spaced values of θ, the same for all c. The implicit function theorem is applicable and leads to a solution $\theta = \theta(r)$ for each initial value and each c. There are thus infinitely many vertical trajectories from w_0, and they are tangent to $2k - 2$ equally spaced directions.

The results apply equally to $w_0 = \infty$, provided that the order at ∞ is defined as the order of $\varphi(1/w) \, d(1/w)^2$ at 0. For instance, $P \, dw^2/w^2$ has a simple pole at ∞, and thus there is a single vertical trajectory which tends to ∞.

For further information on the trajectories we prove a lemma which is essentially a special case of the Gauss-Bonnet formula. In our terminology a *geodesic polygon* II shall be a closed curve consisting of finitely many segments of trajectories which together form the boundary of a simply connected region. We denote the vertices by w_i, and the order of φ at w_i by m_i. Moreover, ω_i shall be the inner angle at w_i, counted so that $0 \leq \omega_i \leq 2\pi$. We agree that w_i counts as a vertex if either $m_i \neq 0$ or $\omega_i \neq \pi$, and we do not need to rule out the possibility of vertices and sides that are superimposed on each other. The polygon will be traversed in the positive direction with respect to the region that it bounds.

Lemma 8-1 The difference N between the number of zeros and poles of φ inside a geodesic polygon is given by

$$N + 2 = \sum_i \left[1 - \frac{(m_i + 2)\omega_i}{2\pi} \right], \qquad (8\text{-}10)$$

where the sum is over the inner angles.

For the proof the argument principle yields

$$\int_{\text{II}} d \arg \varphi = 2\pi N + \sum_i m_i\omega_i,$$

where the change of argument refers only to the sides, not to the vertices. Because the sides are trajectories, $d \arg \varphi + 2\, d(\arg dw) = 0$ along II, and because the tangent changes its direction by 2π,

$$\int_{\mathrm{II}} d\, (\arg dw) + \sum_i (\pi - \omega_i) = 2\pi.$$

Taken together these relations yield (8-10).

REMARK We proved the lemma for a bounded region, but if correctly defined both sides of (8-10) are invariant under linear transformations so that the result remains valid if ∞ is a vertex. We shall not allow ∞ as an interior point of a geodesic polygon.

The quadratic differential $P\, dw^2/w^2$ has $n - 2$ zeros and $n + 1$ coinciding poles at the origin (and a simple pole at ∞). A geodesic polygon encloses either none or all of the poles at the origin. Hence either $N \geq 0$ or $N \leq (n - 2) - (n + 1) = -3$ so that $N + 2$ is never 0 or 1. The same is true for $Q\, dz^2/z^2$ provided that II is contained in $|z| \leq 1$, for Q has at most $n - 2$ zeros in $|z| < 1$.

Lemma 8-2 If a geodesic polygon for $P\, dw^2/w^2$ does not pass through the origin, it has either a vertex $w_i \neq 0, \infty$ with an angle different from $2\pi/(m_i + 2)$, or an angle $< 2\pi$ at ∞. The same is true for $Q\, dz^2/z^2$ provided that the polygon is contained in $|z| \leq 1$.

Indeed, since $N + 2 \neq 0$, one of the terms on the right-hand side of (8-10) must be different from zero. If no term is from a vertex at 0, this can occur only by having one $\omega_i \neq 2\pi/(m_i + 2)$. Note that $m_i = -1$ at ∞, so that in this case the inequality means that $\omega_i < 2\pi$.

A regular trajectory is one that does not pass through a singularity. Every regular trajectory is contained in a maximal regular trajectory $w = \gamma(t)$, $a < t < b$. This includes the case of a closed trajectory, the maximal trajectory being periodic. As t tends to a or b, the point $\gamma(t)$ can either oscillate or tend to a singularity, but it cannot converge to a regular point.

Lemma 8-3 In the case of $P\, dw^2/w^2$ or $Q\, dz^2/z^2$ every maximal regular trajectory tends to a singularity. In particular, there are no closed regular trajectories.

PROOF We consider only the first case, the second being similar. Given any nonsingular point w_0 we shall show that $\gamma(t)$ stays away from w_0 when t is close to a or b. Once this is shown, the lemma follows by an obvious compactness argument.

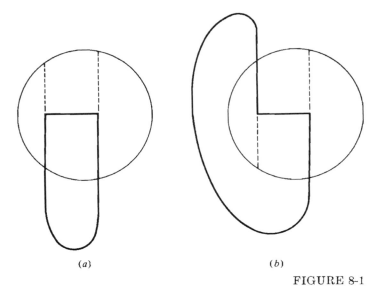

(a)   (b)

FIGURE 8-1

We may assume that the trajectory is a vertical trajectory. In terms of the variable ζ introduced by (8-8) let V be a small neighborhood $|\zeta| < \delta$. If $\gamma(t)$ does not stay away from w_0, the trajectory would intersect V along infinitely many arcs, each represented by a vertical line segment in the ζ plane. Actually, there can be only one such segment. Indeed, if there are two such arcs in succession, the schematic diagram (Fig. 8-1a and b) shows how to construct a geodesic polygon with two right angles or with angles $\pi/2$ and $3\pi/2$ (a closed regular geodesic may be viewed as a degenerate version of the second case). Lemma 8-1 gives $N + 2 = 1$ in the first case and $N + 2 = 0$ in the second. We have already seen that these values are impossible, and Lemma 8-3 is proved.

8-3   THE Γ STRUCTURES

As we have already remarked, the extremal mapping carries singularities into singularities and vertical trajectories into vertical trajectories. The maximal vertical trajectories that can be distinguished from all others are those that either begin or end at a zero or a simple pole. In contrast, there are infinitely many vertical trajectories issuing from a multiple pole, and we cannot distinguish a vertical trajectory that begins and ends at a multiple pole. For this reason we construct, to begin with in the w plane, the graph consisting of all maximal vertical trajectories of $P\,dw^2/w^2$ which do not begin and end at the origin. Following Schaeffer and Spencer [56] we shall denote this graph by Γ_w. The graph is finite, for according to Lemma 8-3 each trajectory that is part of the graph must lead from a zero to a zero, from a zero to a pole, or from the pole at ∞ to the pole at the origin, and there are only a finite number of trajectories issuing from the zeros and from the pole at ∞. The single trajectory from ∞ is always part of Γ_w. The graph divides the plane into regions which we denote generically by Ω_w.

In the z plane we carry out the same construction for $Q\, dz^2/z^2$, but we shall let Γ_z denote only the part of the graph that is contained in $|z| \leq 1$. We recall that the unit circle is a vertical trajectory and that there is at least one zero on the unit circle. Hence the unit circle is always part of Γ_z. The Ω_z are inside the unit circle.

Our next step is to use Lemmas 8-1 and 8-2 to gain information about the Ω_w and Ω_z.

Lemma 8-4   Each Ω_w (or Ω_z) is the inside of a geodesic polygon whose sides belong to the graph Γ_w (or Γ_z). The angle at a vertex that is a zero of order m_i is $2\pi/(m_i + 2)$. In addition the polygon has either one vertex with angle $2\pi/(n - 1)$, or two vertices with angle zero at the origin. One of the Ω_w has angle 2π at ∞.

PROOF   Consider the outside contour of Ω_w. It is obviously a geodesic polygon with sides belonging to Γ_w. The angle at a finite vertex other than the origin is exactly $2\pi/(m_i + 2)$, and if there is a vertex at ∞, the angle is 2π. By Lemma 8-2 the origin must lie on the outer contour. If Ω_w were not simply connected it would have an inside contour which does not pass through the origin. There would be at least one Ω_w' inside this contour, and the outer contour of Ω_w' would not pass through the origin, contrary to what was shown. Hence all Ω_w are simply connected.

The angles of Ω_w at the origin are of the form $\omega_i = 2k_i\pi/(n - 1)$ with integral $k_i \geq 0$, and the corresponding m_i is $-(n + 1)$. It is clear that $N = 0$ in Lemma 8-1. Hence $\Sigma(k_i + 1) = 2$, which is possible only for one $k_i = 1$ or two $k_i = 0$. This is precisely the statement in the lemma that we are proving. The existence of one Ω_w with a vertex at ∞ is obvious. The rest of the proof applies equally to Ω_z.

A geodesic polygon which is also a Jordan curve will be called a geodesic loop.

Lemma 8-5   Every geodesic loop on Γ_w passes through the origin. The same is true of every geodesic loop on Γ_z which is not the whole unit circle.

PROOF If a loop on Γ_w does not pass through the origin, it divides the extended plane into two regions, one of which does not contain 0. This part of the plane is subdivided into regions Ω_w whose boundaries do not pass through the origin, contrary to Lemma 8-4. Similarly, a geodesic loop on Γ_z, which does not pass through the origin and is not identical to the unit circle, divides the unit disk into two parts, one of which does not contain the origin. The same contradiction is reached.

8-4 REGULARITY AND GLOBAL CORRESPONDENCE

We have yet to prove that the extremal function f is analytic on $|z| = 1$ (except for isolated points). This will be a by-product of a closer study of the correspondence between the regions Ω_w and Ω_z.

We use superscripts to distinguish the various $\Omega_w{}^i$ and $\Omega_z{}^j$. In each $\Omega_w{}^i$ and $\Omega_z{}^j$ we choose some fixed determination of

$$\zeta_i(w) = \int \sqrt{P(w)} \, \frac{dw}{w}$$

$$\lambda_j(z) = \int \sqrt{Q(z)} \, \frac{dz}{z}, \tag{8-11}$$

i.e., we choose a branch of the square root and fix the integration constant. This is possible because the $\Omega_w{}^i$ and $\Omega_z{}^j$ are simply connected.

According to Lemma 8-4 there are two types of regions Ω_w. We shall say that $\Omega_w{}^i$ is of *type 1* if it has one vertex with angle $2\pi/(n-1)$ at the origin, and of *type 2* if it has two zero angles at the origin. The same classification applies to the $\Omega_z{}^j$.

Consider first an $\Omega_z{}^j$ of type 1. It is clear that λ_j extends continuously to the boundary of $\Omega_z{}^j$, except perhaps at the vertices, and that each side of the boundary is mapped on a vertical line segment. It follows from (8-11) that the leading term in the expansion of λ_j at the origin has the form $Az^{(1-n)/2}$. Therefore, the origin is thrown to ∞, and the angle of $\Omega_z{}^j$ at the origin is mapped on an angle π at ∞. It is seen in the same way that λ_j remains continuous at all other vertices, and that all angles are straightened. With this information standard use of the argument principle shows that λ_j determines a one-to-one conformal mapping of $\Omega_z{}^j$ onto a right or left half plane, which we shall denote by $\Lambda_z{}^j$. In the case of a region of type 2 there are two vertices that are thrown to ∞, and the corresponding angles are zero. It easily follows that the image region is a vertical strip. The situation is quite similar in the w plane, and we denote the image of $\Omega_w{}^i$ by $\Lambda_w{}^i$. The main theorem can now be stated.

Theorem 8-1 The regions Ω_z and Ω_w can be matched and reindexed so that the extremal function f maps $\Omega_z{}^i$ onto $\Omega_w{}^i$, both regions being of the same type. The function f can be continued analytically to $|z| = 1$, except for one double pole and a finite number of algebraic singularities.

Globally, f maps the unit disk on a slit region obtained by removing a connected part of Γ_w from the w plane. The slit extends to ∞.

PROOF The notation f^{-1} shall refer to the original f as defined for $|z| < 1$, not to its extension. We show first that $f^{-1}(\Gamma_w) \subset \Gamma_z$. For this purpose, let s be a maximal vertical trajectory contained in Γ_w. Then $f^{-1}(s)$ is either empty, or it is a vertical trajectory. In the latter case it is maximal, for an extension of $f^{-1}(s)$ would be mapped by f on an extension of s. It cannot lead from 0 to 0, for then s would have the same property. We conclude that $f^{-1}(s)$ is contained in Γ_z. As for the vertices of Γ_w, it is trivial that the inverse image of each vertex is either empty or a vertex of Γ_z.

Choose any $z_j \in \Omega_z{}^j$ and let $\Omega_w{}^j$ be the region Ω_w that contains $f(z_j)$. If there were a point $z \in \Omega_z{}^j$ with $f(z) \in \Omega_w{}^i \neq \Omega_w{}^j$, there would also be a point in $\Omega_z{}^j$ with $f(z) \in \Gamma_w$. Since this is impossible, $\Omega_w{}^j$ is unique, and $f(\Omega_z{}^j) \subset \Omega_w{}^j$.

We know further that $\zeta_j'[f(z)]^2 f'(z)^2 = \lambda_j'(z)^2$ in $\Omega_z{}^j$. Integration yields

$$\zeta_j[f(z)] = \pm\lambda_j(z) + c_j, \tag{8-12}$$

where c_j is a constant. We shall use (8-12) to prove that f maps $\Omega_z{}^j$ onto $\Omega_w{}^j$.

Define the map h by $h(\lambda_j) = \pm\lambda_j + c_j$, the sign being as in (8-12). Because λ_j and ζ_j are one to one, $f(\Omega_z{}^j) \subset \Omega_w{}^j$ translates into $h(\Lambda_z{}^j) \subset \Lambda_w{}^j$. This already shows that $\Omega_w{}^j$ is of type 1 if $\Omega_z{}^j$ is of type 1. In this case, denote the boundary of the half plane $\Lambda_z{}^j$ by L. Then $h(L)$ is either the boundary of $\Lambda_w{}^j$ or a parallel to it. In the latter case $\zeta_j^{-1}[h(L)] = f[\lambda_j^{-1}(L)]$ would be free from singularities. This is not so because $\lambda_j^{-1}(L)$ is not singularity-free. It follows that h, and hence f, is onto. In the case of type 2 there are two boundary lines L_1 and L_2 of $\Lambda_z{}^j$. Both $h(L_1)$ and $h(L_2)$ must lie on the boundary of $\Lambda_w{}^j$, and they cannot coincide. Therefore $\Omega_w{}^j$ is of type 2, and the mapping is onto.

Note that formula (8-12) will automatically define a continuous extension of f to the part of the boundary of each $\Omega_z{}^j$ which lies on $|z| = 1$. As usual, the analyticity of f, except at the vertices, follows by the reflection principle.

The full image of $|z| = 1$ is a connected part of Γ_w. Since the image does not pass through the origin, it contains no loop, by Lemma 8-5. The image is therefore a slit, which may be branched (the term "tree-shaped" would perhaps be more descriptive). The image of $|z| < 1$ must be the full complement of the slit. This shows that every Ω_w is the image of some Ω_z. Finally, the slit reaches out to ∞ along the single vertical trajectory from that point, for otherwise the complement would not be simply connected. The unique point on $|z| = 1$ that corresponds to $w = \infty$ is a double pole.

The Schiffer method has yielded a qualitative description of the extremal mapping in a situation of great generality. On the other hand, relatively few quantitative results are within the scope of the method.

8-5 THE CASE $n = 3$

In order to illustrate the preceding discussion we shall make a detailed study of the case $n = 3$. Here the problem is to maximize a function $F(a_2, a_3)$. Our earlier notation (8-5) would be

$$P(w) = A_1 w^{-1} + A_2 w^{-2}$$
$$Q(z) = B_{-2} z^{-2} + B_{-1} z^{-1} + B_0 + B_1 z + B_2 z^2,$$

but this conceals the available information, namely, that $Q(z) \geq 0$ on $|z| = 1$ and that $Q(z)$ has at least one double zero on the unit circle. To bring this out we prefer to write

$$P(w) = \frac{A(w - c)}{w^2}$$

$$Q(z) = \frac{B(z - \omega)^2(z - \beta)(z - 1/\bar{\beta})}{z^2},$$

where $c \neq 0$, $|\omega| = 1$, and $|\beta| \leq 1$. The argument of B is in a certain relation to ω and β, but there is no need to make this explicit.

The extremal function $w = f(z)$ must satisfy

$$\sqrt{A} \int \sqrt{w - c}\, \frac{dw}{w^2} = \pm \sqrt{B} \int (z - \omega) \sqrt{(z - \beta)(z - 1/\bar{\beta})}\, \frac{dz}{z^2} + C.$$

The integrals can be evaluated explicitly, and in principle we can express w at least as an implicit function of z, although there remains the problem of choosing signs and integration constants. However, we do not attach much importance to this part of the problem, which leads to complicated formulas at best.

It is of much greater interest to study the Γ structures and the mapping of the $\Omega_z{}^i$ on the $\Omega_w{}^i$. We begin with Γ_w. There are two essentially different cases, depending on whether the vertical trajectory from ∞ leads to c or to 0.

Case I The vertical trajectory from ∞ leads to c. The two other trajectories from c form $120°$ angles, and they lead to the origin where they meet at an angle of $180°$ (Fig. 8-2). There are two regions $\Omega_w{}^1$ and $\Omega_w{}^2$, each mapped on a half plane by means of a branch of

$$\varsigma(w) = \int \sqrt{P(w)}\, \frac{dw}{w} = \sqrt{A} \int \sqrt{w - c}\, \frac{dw}{w^2}.$$

We choose to map $\Omega_w{}^1$ on a left half plane and $\Omega_w{}^2$ on a right half plane (Fig. 8-3). In either case $w = 0$ corresponds to $\varsigma = \infty$. In the left half plane we mark the point c_1 that corresponds to c. In the right half plane there are two points c_2, c_2' corresponding to c, and we also mark the point ∞' that corresponds to ∞. It lies halfway between c_2 and c_2', for both distances are measured by

$$\int_c^\infty \sqrt{|P(w)|}\, \frac{|dw|}{|w|} \tag{8-13}$$

taken along the trajectory from c to ∞.

The segments (c_2, ∞') and (c_2', ∞') have to be identified, as indicated

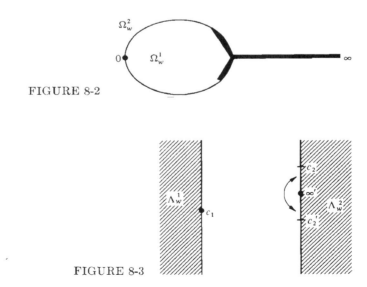

FIGURE 8-2

FIGURE 8-3

by the arrows. It remains to attach the half planes to one another. This is done by identifying the two half lines issuing from c_1 with the half lines starting at c_2 and c_2'. A way to realize this identification is to cut the left half plane along the horizontal line through c_1 and attach each quarter plane to the right half plane (Fig. 8-4). We obtain a conformal model of the w sphere as the complement of a rectangular half strip with two identifications. If desired, the identifications can be made concrete at the expense of allowing folds in the right half plane.

At this point we discover that our model has a line of symmetry. For this reason Fig. 8-2 must also be symmetric with respect to the straight line through 0 and c. In fact, the whole configuration is unique up to a rotation and change of scale. We observe that the width of the half strip is twice the integral (8-13). By residues, it is also equal to 2π times the coefficient of $\log w$ in the development of $\zeta(w)$ at the origin.

For the Γ_z structure there are two possibilities compatible with Case I.

Case Ia $\quad |\beta| < 1$. The Γ_z structure is shown in Fig. 8-5, and the mapping by $\lambda(z)$ is diagrammed in Fig. 8-6. Observe that there is no identification between ω_2 and ω_2'. This segment is the image of the unit circle, and Fig. 8-6 is a model of the unit disk. There is again a line of symmetry, and this symmetry carries over to the z plane.

The extremal mapping is visualized by superimposing Fig. 8-6 on Fig. 8-4. The widths of the half strips must be equal, showing that the coefficient of $\log z$ must be equal to the coefficient of $\log w$. As we pass to $w = f(z)$, we see that $|z| = 1$ is mapped on a proper part of the vertical trajectory from c to ∞. Since the latter is a straight line, we recognize that the only extremal mapping of type Ia is the Koebe mapping.

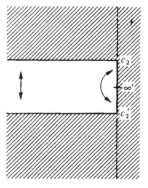

FIGURE 8-4

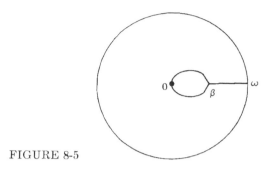

FIGURE 8-5

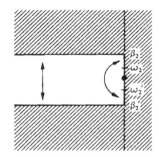

FIGURE 8-6

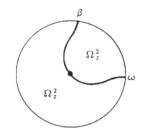

FIGURE 8-7

Case Ib $|\beta| = 1$. The Γ_z structure is shown in Fig. 8-7. Because β and ω are interchangeable, we may assume that the labeling is as in the diagram. In the image half planes (Fig. 8-8) we have marked not only the images of β and ω, but also the points that correspond to c_1, c_2, c_2' and ∞' when the diagram is superimposed on Fig. 8-3. It is readily seen that the distance from β_1 to (c_1) must be equal to the distance from β_2 to (c_2), and the distance from (c_1) to ω_1 is the same as from (c_2') to ω_2. To realize the identifications we cut the left half plane along the horizontal line through (c_1). The quarter planes are attached to the right half plane, but only

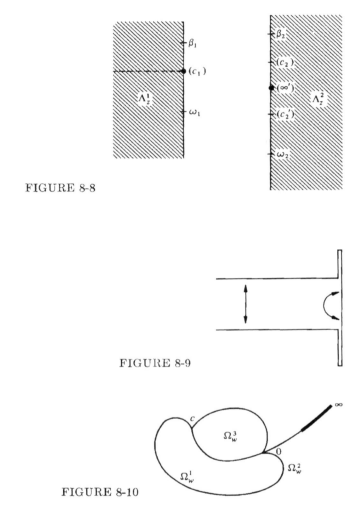

FIGURE 8-8

FIGURE 8-9

FIGURE 8-10

along the half lines above β_1, β_2 and below ω_1, ω_2. Figure 8-9 is an exaggerated picture of the resulting model of the unit disk. The extremal function $f(z)$ maps the unit disk on the complement of a forked slit as indicated by the heavy lines in Fig. 8-2.

Case II The vertical trajectory from ∞ leads to the origin. In the Γ_w structure there are three regions Ω_w, two of type 1 and one of type 2 (Fig. 8-10). The corresponding half planes and strip are shown in Fig. 8-11. We manipulate the strip by cutting it from c_2 to ∞' and rotating the lower part through 180° about ∞', in such a way that identified points will coincide. It is now easy to attach the half planes (Fig. 8-12).

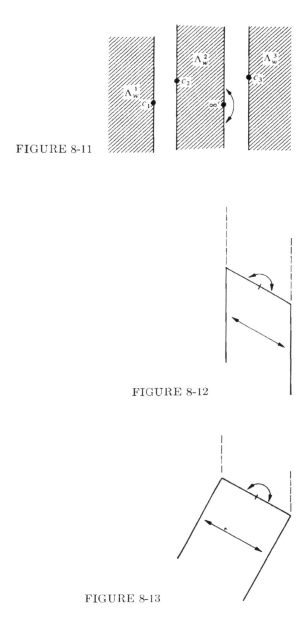

FIGURE 8-11

FIGURE 8-12

FIGURE 8-13

Finally, to make the angles right angles we delete a wedge on the left and attach it on the right (Fig. 8-13). We end up with the same model of the w sphere as in Case I, but now the trajectories are slanted.

There is only one Γ_z structure compatible with Case II (Fig. 8-14). It should be clear by now how to construct the corresponding model of

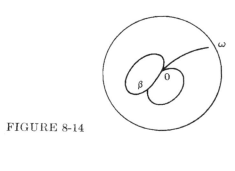

FIGURE 8-14

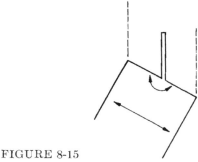

FIGURE 8-15

the unit disk (Fig. 8-15). Figure 8-15 can be superimposed on Fig. 8-13, and we recognize that $f(z)$ maps the unit circle on a slit along the vertical trajectory from ∞ to 0 (the heavy line in Fig. 8-10).

The possible values of (a_2, a_3) form a region in real four-dimensional space, but since (a_2, a_3) can be replaced by $(a_2 e^{i\theta}, a_3 e^{2i\theta})$, we may assume a_3 to be real so that the region is three-dimensional. It may be expected that the coefficient region is bounded by a smooth surface, in which case every point on the boundary surface would correspond to a local extremum of a linear function, for instance, the distance from the tangent plane. If so, we have shown that the boundary point corresponds to a mapping $f(z)$ of the type described under Case I or Case II. Schaeffer and Spencer have shown that this is indeed true, and that the two types of extremal mappings form two boundary surfaces of the coefficient body with only the Koebe mapping as common point. The boundary is of course two-dimensional, the parameters being the lengths $\beta_2 - c_2$ and $c_2' - \omega_2$ in Case I, and the length and angle of the slit in Fig. 8-15 in Case II. The book by Schaeffer and Spencer [56] has a beautiful illustration.

9

RIEMANN SURFACES

9-1 DEFINITION AND EXAMPLES

In the classic literature the term *Riemann surface* is used with two different although related meanings. Riemann, in his thesis, overcomes the difficulty of multiple-valued analytic functions $f(z)$ by means of the suggestive device of letting the variable z vary over a domain which may cover parts of the complex plane several times. Although modern mathematicians frown on the use of multiple-valued functions, the underlying idea is fundamental and can easily be axiomatized. It leads to the topological notion of *covering surface*.

In his later work on the foundations of geometry Riemann introduced what is known as differential manifolds. This idea generalizes in turn to the notion of complex manifolds. In present terminology a Riemann surface is a *one-dimensional complex manifold*.

Definition 9-1 A Riemann surface is a connected Hausdorff space W together with a collection of charts $\{U_\alpha, z_\alpha\}$ with the following properties:

i)   The U_α form an open covering of W.

ii)   Each z_α is a homeomorphic mapping of U_α onto an open subset of the complex plane $\mathbf{C}$.

iii)   If $U_\alpha \cap U_\beta \neq 0$, then $z_{\alpha\beta} = z_\beta \circ z_\alpha^{-1}$ is complex analytic on $z_\alpha(U_\alpha \cap U_\beta)$.

Several comments are in order:

1)   The system $\{U_\alpha, z_\alpha\}$ is said to define a *conformal structure* on W. If it is understood which conformal structure we are referring to, we shall not hesitate to speak of the Riemann surface W.

2)   The topology of W is completely determined by the mappings z_α. Thus, an alternative way would be to require merely that the z_α are one to one and that the sets $z_\alpha(U_\alpha \cap U_\beta)$ are open. The open sets on W are then generated by the inverse images of open sets. The topology is Hausdorff if any two distinct points $p,q \in W$ are either in the same U_α or in disjoint sets U_α and U_β. The connectedness is of course a separate requirement.

3)   A point $p \in U_\alpha$ is uniquely determined by the complex number $z_\alpha(p)$. For this reason z_α is referred to as a *local variable* or *local parameter*. The subscript is frequently dropped, and $z(p)$ is identified with p. For instance, $\Delta_\rho = \{z | z - z_0| < \rho\}$ can refer either to a disk in $\mathbf{C}$, or to its inverse image on W.

4)   The practice of identifying a point on the Riemann surface with the corresponding value of a local variable leads to no difficulty as long as we deal with concepts that are invariant under conformal mapping. Typical instances are the notions of analytic function, harmonic function, subharmonic function, and analytic arc.

5)   It is clear what is meant by a complex analytic mapping from one Riemann surface to another. Two Riemann surfaces are said to be conformally equivalent if there is a one-to-one complex analytic, and hence directly conformal, mapping of one onto the other. They are to be regarded as not essentially different.

6)   Every open connected subset of a Riemann surface is a Riemann surface in its own right.

EXAMPLE 9-1   Let W be the unit sphere $x_1^2 + x_2^2 + x_3^2 = 1$ in three-space. Let U_1 be the complement of $(0,0,1)$ and U_2 the complement of $(0,0,-1)$ with respect to W. As local variables we choose

$$z_1 = \frac{x_1 + ix_2}{1 - x_3}, \qquad z_2 = \frac{x_1 - ix_2}{1 - x_3}$$

on U_1 and U_2, respectively. They are connected by $z_1 z_2 = 1$ on $U_1 \cap U_2$. This choice makes W into a Riemann surface, namely, the Riemann sphere.

EXAMPLE 9-2   Let ω_1 and ω_2 be complex numbers $\neq 0$ whose ratio is not real. We define an equivalence on $\mathbf{C}$ by saying that z_1 and z_2 are equivalent if $z_1 - z_2 = m_1 \omega_1 + m_2 \omega_2$ with integral m_1, m_2. There is a natural projection π which takes $z \in \mathbf{C}$ into its equivalence class $\bar{z}$. We shall define a conformal structure on $T = \pi(\mathbf{C})$. For this purpose let Δ_α be any open set in $\mathbf{C}$ which contains no two equivalent points. We set $U_\alpha = \pi(\Delta_\alpha)$ and define z_α as the inverse of π restricted to Δ_α. These local variables define T as a Riemann surface known as a torus.

9-2 COVERING SURFACES

In Definition 9-1 we required the transition functions $z_{\alpha\beta}$ to be complex analytic. Weaker requirements lead to more general classes of surfaces. We speak of a *surface* as soon as the mappings $z_{\alpha\beta}$ are topological and of a *differentiable surface* if they are of class C^∞. Evidently, a Riemann surface is at the same time a differentiable surface.

Let W and W^* be surfaces, and consider a mapping $f: W^* \to W$. We say that f is a local homeomorphism if every point on W^* has a neighborhood V^* such that the restriction of f to V^* is a homeomorphism. When this is so the pair (W^*,f) is called a *covering surface* of W. The point $f(p^*)$ is the projection of p^*, and p^* is said to lie over $f(p^*)$. It is always possible to choose V^* and $V = f(V^*)$ within the domains of local variables z and z^*. For convenience we again identify V^* and V with their images, and we use the notation $z = f(z^*)$ for the projection map.

Suppose now that W is a Riemann surface with the charts $\{U_\alpha, z_\alpha\}$. Then W^* can be endowed with a unique complex structure which makes the mapping $f: W^* \to W$ complex analytic. Explicitly, this can be done by requiring the charts $\{U_\beta^*, z_\beta^*\}$ to be such that f is one to one on U_β^* and the functions $z_\alpha \circ f \circ z_\beta^{*-1}$ complex analytic whenever they are defined. More informally, the conformal structure on W^* is such that $g \circ f$ is analytic on W^* whenever g is analytic on W.

We remark that when W^* and W are Riemann surfaces it is also possible to consider complex analytic mappings $f: W^* \to W$ which are not necessarily locally one to one. In this case (W^*,f) may be considered a ramified covering surface, and in contrast an ordinary covering surface is said to be smooth. In the following discussion we shall consider only smooth covering surfaces.

Let V be an open set on W. We shall say that V is *evenly covered* by (W^*,f) if every component of the inverse image $f^{-1}(V)$ is in one-to-one correspondence with V. This correspondence is always topological, and if W and W^* are Riemann surfaces, it is a conformal mapping.

Definition 9-2   A covering surface (W^*,f) of W is said to be complete if every point has an evenly covered open neighborhood.

Every connected subset of an evenly covered set is itself evenly covered. For this reason it is sufficient to consider neighborhoods V which are homeomorphic to a disk.

Lemma 9-1   A complete covering surface covers each point the same number of times.

This is an immediate consequence of the definition together with the connectedness of W. It is indeed easy to show that the set of points that are covered exactly n times is open and closed. The number of times each point is covered is called the number of sheets.

Let γ be an arc on W, that is, a continuous mapping $\gamma: [a,b] \to W$. We say that the arc $\gamma^*: [a,b] \to W^*$ covers γ, or that γ can be lifted to γ^*, if $f[\gamma^*(t)] = \gamma(t)$ for all $t \in [a,b]$. The initial point $\gamma^*(a)$ lies over the initial point $\gamma(a)$.

Theorem 9-1   If (W^*,f) is complete, every arc γ can be lifted to a unique γ^* from any initial point p_0^* over p_0, the initial point of γ.

PROOF   Let E be the set of all $\tau \in [a,b]$ such that $\gamma[a,\tau]$ can be uniquely lifted to $\gamma^*[a,\tau]$ with the initial point p^*. E is not empty, for $a \in E$. If $\tau \in E$, we determine an evenly covered neighborhood V of $\gamma(\tau)$ and choose δ so small that $\gamma[\tau,\tau + \delta] \subset V$. The point $\gamma^*(\tau)$ belongs to a component V^* of $f^{-1}(V)$. Since $f: V^* \to V$ is topological, there is a unique way of extending γ^* to $[\tau,\tau + \delta]$. This proves that E is relatively open. A similar proof shows that the complement of E is open. Hence $E = [a,b]$.

9-3   THE FUNDAMENTAL GROUP

Suppose that the arcs $\gamma_1: [0,1] \to W$ and $\gamma_2: [0,1] \to W$ have common end points: $\gamma_1(0) = \gamma_2(0)$, $\gamma_1(1) = \gamma_2(1)$. A continuous mapping γ:

$[0,1] \times [0,1] \to W$ is a deformation of γ_1 into γ_2 if $\gamma(0,u) = \gamma_1(0)$, $\gamma(1,u) = \gamma_1(1)$, $\gamma(t,0) = \gamma_1(t)$, $\gamma(t,1) = \gamma_2(t)$. In other words, there is a continuously changing arc between fixed end points whose initial position is γ_1 and terminal position γ_2. When such a deformation exists, we say that γ_1 is *homotopic* to γ_2, and we write $\gamma_1 \approx \gamma_2$.

We remark that it is only for the sake of convenience that we refer all arcs to the parametric interval $[0,1]$. The general case can be reduced to this by a linear change of parameter.

As an example we note that if W is a convex region in the plane, then any two arcs with common end points are homotopic by way of the deformation $\gamma(t,u) = (1 - u)\gamma_1(t) + u\gamma_2(t)$.

Suppose that we change the parametric representation of an arc from $\gamma(t)$ to $\gamma[\tau(t)]$, where $\tau(t)$ increases monotonically and continuously from 0 to 1. Then these two arcs, which are geometrically identical, are also homotopic as seen from the mapping $\gamma[(1 - u)t + u\tau(t)]$.

The relation $\gamma_1 \approx \gamma_2$ is evidently an equivalence relation. The equivalence classes are called *homotopy classes*, and we denote the homotopy class of γ by $\{\gamma\}$. Two arcs in the same homotopy class have common end points. We have just seen that reparametrization does not alter the homotopy class.

Suppose that γ_2 begins where γ_1 ends. Then the product $\gamma = \gamma_1\gamma_2$ (γ_1 followed by γ_2) is defined by

$$\gamma(t) = \begin{cases} \gamma_1(2t) & \text{for} & t \in [0,\tfrac{1}{2}] \\ \gamma_2(2t - 1) & \text{for} & t \in [\tfrac{1}{2},1]. \end{cases}$$

This construction is compatible with homotopy in the sense that $\gamma_i \approx \gamma_i'$ ($i = 1,2$) implies $\gamma_1\gamma_2 \approx \gamma_1'\gamma_2'$. In view of this property we can define a product of two homotopy classes by setting $\{\gamma_1\}\{\gamma_2\} = \{\gamma_1\gamma_2\}$.

The product suffers from the disadvantage that it is not always defined. To eliminate this difficulty we pick a point $p_0 \in W$ and consider only arcs that begin and end at p_0. With this restriction all homotopy classes can be multiplied. Moreover, the homotopy class of the degenerate arc given by $\gamma(t) = p_0$ acts as a multiplicative unit and is therefore denoted by **1** (it depends on p_0, but there is little need for more precise notation). An easy verification shows that multiplication is associative. Also, every homotopy class has an inverse, for $\{\gamma\}\{\gamma^{-1}\} = 1$, where γ^{-1} is γ traced backward, i.e., $\gamma^{-1}(t) = \gamma(1 - t)$. These properties show that the homotopy classes of closed curves from p_0 form a group. It is called the *fundamental group* of W with respect to p_0, and the standard notation is $\pi_1(W,p_0)$.

What happens if we replace p_0 by another point p_1? Since W is connected, we can draw an arc σ from p_0 to p_1. If γ_1 begins and ends at p_1, then $\gamma = \sigma\gamma_1\sigma^{-1}$ begins and ends at p_0, and $\{\gamma\}$ depends only on $\{\gamma_1\}$. This correspondence between $\{\gamma\}$ and $\{\gamma_1\}$ is one to one, for $\gamma \approx \sigma\gamma_1\sigma^{-1}$ if and only if $\gamma_1 \approx \sigma^{-1}\gamma\sigma$. Moreover, the correspondence preserves products, and hence is an isomorphism between $\pi_1(W,p_0)$ and $\pi_1(W,p_1)$. In other words, the choice of p_0 makes very little difference. As an abstract group the fundamental group is denoted by $\pi_1(W)$.

A surface W is said to be *simply connected* if $\pi_1(W)$ reduces to the unit element. We recall that a plane region is simply connected if and only if its complement with respect to the extended plane is connected.

9-4 SUBGROUPS AND COVERING SURFACES

From now on all covering surfaces are understood to be complete. In this section we study the relations between the fundamental group and the covering surfaces of a surface W.

If (W_1,f_1) is a covering surface of W, and if (W_2,f_{21}) is a covering surface of W_1, then $(W_2,f_1 \circ f_{21})$ is a covering surface of W. We say in this situation that the latter is a stronger covering surface. More symmetrically, (W_2,f_2) is stronger than (W_1,f_1) if there exists a mapping f_{21} such that $f_2 = f_1 \circ f_{21}$ and (W_2,f_{21}) is a covering surface of W_1. This relationship is transitive and defines a partial ordering. If two covering surfaces are mutually stronger than the other, they are equivalent and we regard them as essentially the same.

Consider (W^*,f) over W. We choose $p_0 \in W$ and p_0^* over p_0. Let γ be a closed curve on W from p_0, and let γ^* be the lifted arc with initial point p_0^*. Then γ^* may or may not be closed. The *monodromy theorem* states that γ^* is homotopic to 1, and consequently closed, if γ is homotopic to 1 (for a proof see C.A., p. 285). Therefore, whether γ^* is closed or not depends only on the homotopy class of γ.

Let D be the set of all homotopy classes $\{\gamma\}$ such that γ^* is closed. If $\{\gamma\}$ belongs to D, so does $\{\gamma^{-1}\}$, and if $\{\gamma_1\}$ and $\{\gamma_2\}$ are in D, so is $\{\gamma_1\gamma_2\}$. Hence D is a subgroup of $\pi_1(W,p_0)$. Observe that D depends on the choice of p_0^*.

The dependence on p_0^* is quite easy to determine. Suppose that we replace p_0^* by p_1^* with the same projection p_0. Let σ^* be an arc from p_0^* to p_1^*. Its projection σ is a closed curve from p_0. If γ is a closed curve from p_0, it is readily seen that $\sigma\gamma\sigma^{-1}$ lifts to a closed curve from p_0^* if and only if γ lifts to a closed curve from p_1^*. If D_1 corresponds to p_1^* in the same way that D corresponds to p_0^*, we thus have $D_1 = \{\sigma\}^{-1}D\{\sigma\}$. In other words, D and D_1 are conjugate subgroups. Conversely, every conjugate subgroup of D can be obtained in this manner.

Theorem 9-2   The construction that we have described determines a one-to-one correspondence between the classes of conjugate subgroups of $\pi_1(W,p_0)$ and the equivalence classes of covering surfaces (W^*,f). Moreover, if D and W^* correspond to each other, then $\pi_1(W^*)$ is isomorphic to D.

PROOF   We have already shown that every (W^*,f) determines a class of conjugate subgroups, and it is evident that equivalent covering surfaces determine the same class.

Conversely, if we start from a subgroup D, we have to construct a corresponding surface W^*. Let σ_1 and σ_2 be arcs on W from p_0. We shall write $\sigma_1 \sim \sigma_2$ if they have the same end point, and if $\sigma_1\sigma_2^{-1} \in D$. This is obviously an equivalence relation. The points of W^* will be equivalence classes of arcs σ, the class of σ being denoted by $[\sigma]$. The projection map f will take $[\sigma]$ into the common terminal point of all $\sigma \in [\sigma]$.

It is not difficult to provide W^* with a surface structure. To describe the process in some detail, let the structure of W be given by charts $\{U_\alpha, z_\alpha\}$, where the U_α are topological disks. Choose a point $q_0 \in U_\alpha$ and $q_0^* = [\sigma_0]$ over q_0. For any $q \in U_\alpha$ we draw an arc σ from q_0 to q within U_α. Then $\sigma_0\sigma$ determines a point $[\sigma_0\sigma]$ which depends on q_0^* and q, but not on the choice of σ. We have obtained a set $U_\alpha^*(q_0^*) \subset W^*$ which is in one-to-one correspondence with its projection U_α, and we define the structure of W^* by means of the charts $\{U_\alpha^*(q_0^*), z_\alpha \circ f\}$.

It is necessary to verify that the induced topology is Hausdorff. Let p_1^* and p_2^* be distinct points of W^*. If they have different projections, the existence of disjoint neighborhoods is obvious. Assume now that $f(p_1^*) = f(p_2^*)$ and $p_1^* \in U_\alpha^*(q_0^*)$, $p_2^* \in U_\beta^*(q_1^*)$. We can write $p_1^* = [\sigma_0\sigma']$, $p_2^* = [\sigma_1\sigma'']$. By assumption, $\sigma_0\sigma'\sigma''^{-1}\sigma_1^{-1}$ is a closed curve, but its homotopy class is not in D. If $U_\alpha^*(q_0^*)$ and $U_\beta^*(q_1^*)$ had a common point, it would have two representations $[\sigma_0\tau'] = [\sigma_1\tau'']$, and $\sigma_0\tau'\tau''^{-1}\sigma_1^{-1}$ would belong to D. But if the terminal points of τ' and σ' are in the same component of $U_\alpha \cap U_\beta$, it is easy to see that $\tau'\tau''^{-1} \approx \sigma'\sigma''^{-1}$, and we would arrive at a contradiction. It follows that p_1^* and p_2^* have disjoint neighborhoods.

Our proof also shows that any two $U_\alpha^*(q_0^*)$ with the same U_α are either disjoint or identical. This in turn implies that the $U_\alpha^*(q_0^*)$ are components of $f^{-1}(U_\alpha)$, and hence (W^*,f) is a complete covering surface.

We now show that (W^*,f) determines the subgroup D or one of its conjugates. As initial point we choose $p_0^* = [1]$, the equivalence class of the constant curve from p_0. Let σ be a closed curve from p_0, parametrized over $[0,1]$, and denote its restriction to $[0,\tau]$ by σ_τ. Then the lifted arc $\tilde{\sigma}$ is given by $\tilde{\sigma}(\tau) = [\sigma_\tau]$. It is closed if and only if $\{\sigma\} \in D$, which is what we wanted to prove.

It remains to prove that $\pi_1(W^*)$ is isomorphic to D. Let γ^* be a closed curve on W^* from p_0^*, and let γ be its projection. Then $\{\gamma\} \in D$, and if γ_1^* and γ_2^* are homotopic, so are their projections, for a deformation on W^* projects to a deformation on W. Hence projection induces a mapping of $\pi_1(W^*, p_0^*)$ into D. This mapping is onto by the definition of D. It is obviously product preserving, and it is one to one, for $\gamma \approx 1$ implies $\gamma^* \approx 1$ by the monodromy theorem. We conclude that the projection map defines an isomorphism between $\pi_1(W^*, p_0^*)$ and D.

In the applications of Theorem 9-2 there are two extreme cases. First, if D is the whole group $\pi_1(W, p_0)$, two arcs from p_0 are equivalent as soon as they lead to the same point, and the projection f is a homeomorphism so that W^* can be identified with W. The other extreme occurs when D reduces to the unit element. The corresponding covering surface is called the *universal covering surface* of W, and we denote it by $\tilde{W}$. It has the property that an arc on $\tilde{W}$ is a closed curve if and only if its projection is homotopic to 1. Furthermore, $\pi_1(\tilde{W}) = 1$, so that $\tilde{W}$ is simply connected.

The subgroups of the fundamental group have the same partial ordering as the corresponding covering surfaces. To be more specific, suppose that D_1 and D_2 correspond to W_1^* and W_2^*, respectively. If $D_1 \subset D_2$, then W_1^* is stronger than W_2^*. Conversely, if W_1^* is stronger than W_2^*, then D_2 contains a conjugate of D_1. In particular, the universal covering surface is the strongest. The proofs are left to the reader.

9-5 COVER TRANSFORMATIONS

Given a covering surface (W^*, f) of W, let φ be a homeomorphic mapping of W^* onto itself. It is called a *cover transformation* of W^* over W if $f \circ \varphi = f$, that is to say, if p and $\varphi(p)$ have the same projection. If W and W^* are Riemann surfaces, then φ is a conformal mapping. Indeed, the local variables on W and W^* may be chosen as z_α and $z_\alpha^* = z_\alpha \circ f$. To say that φ is conformal is to say that $z_\beta^* \circ \varphi \circ z_\alpha^{*-1}$ is conformal where it is defined. But $z_\beta^* \circ \varphi \circ z_\alpha^{*-1} = z_\beta \circ f \circ \varphi \circ f^{-1} \circ z_\alpha^{-1} = z_\beta \circ z_\alpha^{-1}$, and this is conformal by hypothesis.

Theorem 9-3 A cover transformation, other than the identity, has no fixed points.

PROOF   Suppose that $\varphi(p_0^*) = p_0^*$. By the definition of covering surface p_0^* has a neighborhood V^* such that $f \colon V^* \to f(V^*)$ is a homeomorphism. Let $U^* \subset V^*$ be a neighborhood of p_0^* such that $\varphi(U^*) \subset V^*$. If $p^* \in U^*$, we have $f[\varphi(p^*)] = f(p^*) \in f(V^*)$. Since p^* and $\varphi(p^*)$ are both in V^*, it follows that $\varphi(p^*) = p^*$. We conclude that the set of fixed points is open. It is trivially closed, and the theorem follows from the connectedness of W^*.

The cover transformations of (W^*, f) over W form a group. We shall show that there is a simple connection between the group of cover transformations and the subgroup D that corresponds to (W^*, f) in the sense of Theorem 9-2.

Theorem 9-4   The group of cover transformations of (W^*, f) over W is isomorphic to $N(D)/D$, where $N(D)$ is the normalizer of D in $\pi_1(W, p_0)$.

PROOF   Recall that $g \in \pi_1(W, p_0)$ belongs to $N(D)$ if and only if $gD = Dg$. Consider a closed curve γ from p_0 such that $\{\gamma\} \in N(D)$. With this γ we associate a mapping φ_γ as follows. Join p_0^* to p^* by an arc σ^* with projection σ and define $\varphi(p^*)$ as the end point of the lifted arc $(\gamma\sigma)^*$. It must be shown that the result does not depend on the choice of σ^*. Suppose that σ^* is replaced by σ_1^*. Then $\{\sigma\sigma_1^{-1}\} \in D$, and hence $\{\gamma\sigma\sigma_1^{-1}\gamma^{-1}\} \in D$. Therefore, $(\gamma\sigma)^*$ has the same end point as $(\gamma\sigma_1)^*$. It is quite obvious that φ_γ is a cover transformation, and that $\varphi_{\gamma\gamma'} = \varphi_\gamma \circ \varphi_{\gamma'}$. We observe further that φ_γ is the identity if and only if $\{\gamma\} \in D$. Hence our construction defines an isomorphic mapping of $N(D)/D$ into the group of cover transformations.

Conversely, let φ be a cover transformation. Let γ^* be an arc from p_0^* to $\varphi(p_0^*)$ with projection γ. Then $\varphi_\gamma(p_0^*) = \varphi(p_0^*)$ so that $\varphi_\gamma\varphi^{-1}$ has the fixed point p_0^*. Hence $\varphi = \varphi_\gamma$ by Theorem 9-3, and we have shown that every cover transformation is of the form φ_γ. The proof is complete.

A particularly important case occurs when $N(D) = \pi_1(W, p_0)$, that is, for a normal subgroup D. There is then a φ_γ corresponding to every closed curve γ from p_0, and the cover transformations are transitive in the sense that there is a unique φ which sends p_0^* to any point over p_0. A covering with this property is said to be *regular*, and it is a property that does not depend on the choice of p_0. On a regular covering surface points with the same projection are indistinguishable from each other.

9-6 SIMPLY CONNECTED SURFACES

We have defined a simply connected surface as one whose fundamental group reduces to the unit element. In view of Theorem 9-2 this means that every covering surface has a single sheet. The consequences of this property are somewhat indirect, for in most applications it is necessary to construct a covering surface where none was originally present. The construction is illustrated in the proof of the following fairly general theorem.

> **Theorem 9-5** Let W be a simply connected surface and let $\{U_\alpha\}$ be a covering of W by open connected sets. On each U_α let there be given a family Φ_α of functions such that the following conditions are satisfied:
>
> a) If $\varphi_\alpha \in \Phi_\alpha$, $\varphi_\beta \in \Phi_\beta$, and if $V_{\alpha\beta}$ is a component of $U_\alpha \cap U_\beta$, then $\varphi_\alpha(p) = \varphi_\beta(p)$ either for all $p \in V_{\alpha\beta}$ or for no such p.
> b) If $\varphi_\alpha \in \Phi_\alpha$, and $V_{\alpha\beta}$ is a component of $U_\alpha \cap U_\beta$, then there exists a $\varphi_\beta \in \Phi_\beta$ such that $\varphi_\alpha = \varphi_\beta$ in $V_{\alpha\beta}$.
>
> In these circumstances there exists a function φ on W whose restriction to any U_α belongs to Φ_α. Moreover, φ is uniquely determined by its restriction to a single U_α.

REMARK   We have deliberately failed to specify the nature of the functions $\varphi_\alpha \in \Phi_\alpha$ as being completely irrelevant. Actually, they are best thought of as nametags. A more abstract formulation would have its advantages, but we have chosen one that comes as close as possible to the most common applications.

PROOF   Consider all pairs (p,φ) such that $p \in U_\alpha$ and $\varphi \in \Phi_\alpha$ for some α. The relation $(p,\varphi) \sim (q,\psi)$, if $p = q$ and $\varphi(p) = \psi(q)$, is an equivalence relation. We denote the equivalence class of (p,φ) by $[p,\varphi]$. Let W^* be the set of all such equivalence classes, and let f denote the function that maps $[p,\varphi]$ on p.

For a given α and $\varphi_\alpha \in \Phi_\alpha$ let $U^*[\alpha,\varphi_\alpha]$ denote the set of all $[p,\varphi_\alpha]$ with $p \in U_\alpha$. The map f sets up a one-to-one correspondence between $U^*[\alpha,\varphi_\alpha]$ and U_α. This correspondence induces a topology on W^*, and it is a consequence of (a) that the topology is Hausdorff.

Let W_0^* be a component of W^*. We contend that (W_0^*,f) is a complete covering surface of W. To see this consider $p^* = [p,\varphi] \in f^{-1}(U_\alpha)$. Then $\varphi \in \Phi_\beta$ and $p \in U_\alpha \cap U_\beta$ for some β. By (b) there exists a $\psi \in \Phi_\alpha$ such that $\varphi(p) = \psi(p)$, from which it follows that $p^* \in U^*[\alpha,\psi]$. On the other hand, each $U^*[\alpha,\varphi_\alpha]$ is contained in $f^{-1}(U_\alpha)$, and we have the representation

$$f^{-1}(U_\alpha) = \bigcup_{\varphi_\alpha \in \Phi_\alpha} U^*[\alpha,\varphi_\alpha].$$

Here each $U^*[\alpha,\varphi_\alpha]$ is open and connected, and by virtue of (a) the sets that correspond to different φ_α are either identical or disjoint. Hence the $U^*[\alpha,\varphi_\alpha]$ are the components of $f^{-1}(U_\alpha)$, and those that belong to W_0^* are the components of $f^{-1}(U_\alpha) \cap W_0^*$. Since they are in one-to-one correspondence with U_α, we have shown that (W_0^*,f) is a complete covering surface of W.

Our assumption was that W is simply connected. Hence $f: W_0^* \to W$ has an inverse, and $f^{-1}(U_\alpha) = U^*[\alpha,\varphi_\alpha]$ for a $\varphi_\alpha \in \Phi_\alpha$. If $U_\alpha \cap U_\beta \neq 0$, the corresponding $\varphi_\alpha,\varphi_\beta$ coincide on $U_\alpha \cap U_\beta$, and together they define a global function φ. To make φ coincide with a given φ_{α_0} on U_{α_0} all that is needed is to choose W_0^* as the component of W^* which contains $U^*[\alpha_0,\varphi_{\alpha_0}]$.

Corollary 9-1 Assume that the complex-valued function F is continuous and $\neq 0$ on a simply connected surface W. Then it is possible to define a continuous function f on W such that $e^f = F$.

PROOF Every p_0 has an open connected neighborhood in which $|F(p) - F(p_0)| < |F(p_0)|$. Let the system U_α consist of all such neighborhoods, and define $(\log F)_\alpha$ as a single-valued continuous branch of $\log F$ in U_α. The family Φ_α will consist of all functions $(\log F)_\alpha + n2\pi i$ with integral n. Conditions (a) and (b) are trivially fulfilled. Hence there exists a function f which is equal to some $(\log F)_\alpha + n2\pi i$ in each U_α. This function is continuous and satisfies $e^f = F$.

Corollary 9-2 Let u be a harmonic function on a simply connected surface W. Then u has a conjugate harmonic function on W.

PROOF Choose the U_α conformally equivalent to a disk. Then u has a conjugate harmonic function v_α in each U_α. Let Φ_α consist of all functions $v_\alpha + c$ with constant c. The theorem permits us to find a global function v which is of the form $v_\alpha + c$ on each U_α.

NOTES Riemann's ideas were profound, but vaguely expressed. It seems that Klein was the first to understand conformal structure in its modern sense, although in a very informal setting. The present concept of Riemann surface, and its generalization to complex manifolds, goes back to Weyl's monumental "Die Idee der Riemannschen Fläche" [66].

The reader is also advised to consult Springer [60] for a less arduous approach and Ahlfors and Sario [5] for more detail.

10

THE UNIFORMIZATION THEOREM

10-1 EXISTENCE OF THE GREEN'S FUNCTION

In this chapter we shall prove the famous *uniformization theorem* of Koebe. This is perhaps the single most important theorem in the whole theory of analytic functions of one variable. It does for Riemann surfaces what Riemann's mapping theorem does for plane regions. As a matter of fact, as soon as the uniformization theorem is proved, it is not necessary to consider Riemann surfaces more general than the disk, the plane, and the sphere. It must be admitted, of course, that the reduction to these cases does not always simplify matters.

The early proofs of the uniformization theorem were long and unperspicuous. By taking advantage of all simplifications that are now available the proof can be reduced to manageable proportions. The only constructive element is contained in the Perron method for solving the Dirichlet problem. In addition, the proof rests on repeated applications of the maximum principle. In the final stage we shall make use of a special argument due to Heins which in one step eliminates several difficulties in the classic proofs.

We shall first discuss the existence of Green's function on a Riemann surface. As already indicated, the discussion is based on Perron's method, which uses subharmonic functions. We recall that subharmonicity is invariant under conformal mappings. It is for this reason that it is possible to consider subharmonic functions on Riemann surfaces.

It is customary to require subharmonic functions to be upper semicontinuous and to allow $-\infty$ as a value. For our purposes, however, it is sufficient to consider only continuous subharmonic functions, except that we include functions which tend to $-\infty$ at isolated points.

Let W be a Riemann surface. A *Perron family* is a family V of subharmonic functions on W subject to the following conditions:

i)   If v_1 and v_2 are in V, so is max (v_1, v_2).

ii)   Let Δ be a Jordan region on W. Suppose that $v \in V$ and let $\bar{v}$ be a function which is harmonic in Δ with the same boundary values as v, and which agrees with v on the complement of Δ. It is well known that $\bar{v}$ is subharmonic. For a Perron family we require that $\bar{v} \in V$.

Observe that $\bar{v}$ always exists and can be constructed by means of the Poisson integral.

The basic property of Perron families is the following:

Theorem 10-1   If V is a Perron family, the function u defined by $u(p) = \sup v(p)$ for $v \in V$ is either harmonic or identically $+\infty$.

For plane regions the theorem is proved in C.A., pp. 240–241, and the proof easily generalizes to Riemann surfaces.

Consider a point $p_0 \in W$ and let z be a local variable at p_0 with $z(p_0) = 0$. Let V_{p_0} be the family of functions v with the following properties:

a) v is defined and subharmonic on $W - \{p_0\}$.

b) v is identically zero outside a compact set.

c) $\overline{\lim_{p \to p_0}} [v(p) + \log |z(p)|] < \infty$.

It is quite evident that V_{p_0} is a Perron family. If $\sup v$ is finite, and hence harmonic, we say that W has a Green's function with a pole at p_0, and we denote $\sup v(p)$ by $g(p, p_0)$. The Green's function does not depend on the choice of local variable $z(p)$ at p_0, for condition (c) is clearly independent of that choice. We shall see later that the existence or non-existence of g is also independent of p_0.

Suppose that the disk $|z| \leq r_0$ is contained in the range of $z(p)$. Set

$v_0(p) = \log r_0 - \log |z(p)|$ when $|z(p)| \leq r_0$ and $v_0(p) = 0$ everywhere else on W. Then $v_0 \in V_{p_0}$ so that $g(p,p_0) \geq v_0(p)$. We conclude above all that $g(p,p_0) \to \infty$ when $p \to p_0$. In particular, $g(p,p_0)$ is not constant.

If W is compact, the Green's function cannot exist, for in that case $g(p,p_0)$ would have a minimum, and this cannot happen since $g(p,p_0)$ is not a constant and harmonic everywhere except at p_0.

We list the following important properties of the Green's function:

AI $g(p,p_0) > 0$.

AII inf $g(p,p_0) = 0$.

$AIII$ $g(p,p_0) + \log |z(p)|$ has a harmonic extension to a neighborhood of p_0.

The first property follows from the fact that 0 is in V_{p_0}. The others are not as obvious and will be proved later.

10-2 HARMONIC MEASURE AND THE MAXIMUM PRINCIPLE

A noncompact surface is also said to be *open*. It can be compactified by adding a single point "at infinity" whose neighborhoods are the sets with compact complement. In the theory of Riemann surfaces the added point is also referred to as the *ideal boundary*. A sequence $\{p_n\}$ converges to ∞, or to the ideal boundary, if p_n lies outside any given compact set for all sufficiently large n.

Let W be an open Riemann surface and let K be a compact set whose complement $W - K$ is connected. We introduce a Perron family V_K as follows:

i) $v \in V_K$ is defined and subharmonic on $W - K$.

ii) $v \in V_K$ is ≤ 1 on $W - K$.

iii) If $v \subset V_K$, then $\overline{\lim} v(p_n) \leq 0$ when $p_n \to \infty$.

A more precise version of (iii) would read: Given any $\epsilon > 0$ there exists a compact set K_ϵ such that $v(p) < \epsilon$ when $p \in W - K_\epsilon$.

It is immediately evident that V_K is a Perron family. Because the $v \in V_K$ are uniformly bounded, the harmonic function $u_K = \sup v$ will always exist, and $0 \leq u_K \leq 1$. It may happen that $u_K = 0$ or $u_K = 1$.

We can rule out the first possibility by insisting that K have an interior point. To see this, let K^0 be the interior of K and let p_0 be a boundary point of K^0. Suppose that a local variable z maps a neighborhood of p_0 on the disk $|z| < 1$. Then there exist concentric disks $|z - z_0| < \delta$ and $|z - z_0| < 2\delta$, both contained in $|z| < 1$, such that the smaller disk is contained in $z(K)$ while the larger is not. We define a function v by the following rule: (1) If $z(p)$ is defined and satisfies $\delta < |z(p) - z_0| < 2\delta$, we set

$$v(p) = \log \frac{2\delta}{|z(p) - z_0|} : \log 2;$$

(2) otherwise we set $v(p) = 0$. The restriction of v to $W - K$ belongs to V_K and is not identically zero. Hence $u_K > 0$.

There remain two possibilities: either $0 < u_K < 1$ or $u_K = 1$. In the first case we call u_K the *harmonic measure* of K; in the second case we say that the harmonic measure does not exist. We are going to show that these alternatives do not depend on K, but only on the surface W. It is therefore natural to say that the existence of the harmonic measure is a property of the ideal boundary.

There is one more important property relating to a compact set K on W. Let u be harmonic and bounded above on $W - K$. We say that the *maximum principle is valid* on $W - K$ if $\overline{\lim_{p \to K}} u(p) \leq 0$ implies $u \leq 0$ on $W - K$.

Observe that we would not expect this to hold unless u is known to be bounded, that is, to satisfy some inequality of the form $u \leq M$. Again, we shall find that the validity of the maximum principle depends only on W and not on K. For the maximum principle it is not necessary to assume that K has interior points.

10-3 EQUIVALENCE OF THE BASIC CONDITIONS

We turn now to the main theorem linking the concepts introduced above.

Theorem 10-2 The following properties of an open Riemann surface are equivalent:

$\quad i)$ Green's function exists.
$\quad ii)$ Harmonic measure exists.
$\quad iii)$ The maximum principle is not valid.

More precisely, we shall label these statements $(i)_{p_0}$, $(ii)_K$, $(iii)_K$ when they refer to a specific p_0 or K. We claim that all statements are simultaneously true or false, regardless of the choice of p_0 or K.

It is evidently sufficient to establish the following implications:

I $(i)_{p_0} \Rightarrow (iii)_K$ if $p_0 \in K$.

II $(ii)_K \Rightarrow (i)_{p_0}$ if $p_0 \in K^0$ ($=$ int K).

III $(iii)_K \Rightarrow (ii)_{K'}$ for all K and K'.

PROOF OF I   The function $-g(p,p_0)$ has a maximum m on K. It is bounded above by 0 on $W - K$. If the maximum principle were valid in $W - K$, we would have $-g(p,p_0) \leq m$ in $W - K$. Hence m would be the maximum value of $-g(p,p_0)$ in the whole plane, and it would be assumed at an interior point of the complement of $\{p_0\}$. This contradicts the classic maximum principle.

PROOF OF II   We choose a neighborhood of p_0, contained in K, which is conformally equivalent to $|z| < 1$ with p_0 corresponding to $z = 0$. Let K_1 and K_2 be the sets on W that correspond to the closed disks $|z| \leq r_1$ and $|z| \leq r_2$ with $0 < r_1 < r_2 < 1$, and denote their boundaries by ∂K_1, ∂K_2. If u_K exists, as we assume, so does u_{K_1}. Consider $v \in V_{p_0}$ and replace it by $v^+ = \max (v,0)$ which is also in V_{p_0}. The inequality

$$v^+(p) \leq (\max_{K_1} v^+) u_{K_1}(p)$$

holds near the ideal boundary and also on ∂K_1. Hence it holds outside of K_1, and we obtain in particular

$$\max_{\partial K_2} v^+ \leq (\max_{\partial K_1} v^+)(\max_{\partial K_2} u_{K_1}). \tag{10-1}$$

Next we consider the function $v^+(p) + (1 + \epsilon) \log |z(p)|$ on K_2 with $\epsilon > 0$. It becomes negatively infinite when $p \to p_0$. Therefore, its maximum is taken on ∂K_2, and we obtain

$$\max_{\partial K_1} v^+ + (1 + \epsilon) \log r_1 \leq \max_{\partial K_2} v^+ + (1 + \epsilon) \log r_2. \tag{10-2}$$

On combining (10-1) with (10-2) and letting ϵ become zero we find

$$\max_{\partial K_1} v^+ \leq (1 - \max_{\partial K_2} u_{K_1})^{-1} \log \frac{r_2}{r_1},$$

where it is known that $\max_{\partial K_2} u_{K_1} < 1$. It follows that v^+, and consequently v, is uniformly bounded above on ∂K_1. Hence $g(p,p_0)$ exists.

PROOF OF III   We show that the nonexistence of $u_{K'}$ implies that the maximum principle is valid in $W - K$.

Assume first that $K' \subset K$. Let u be harmonic outside of K with $u \leq 1$ and $\overline{\lim}\, u(p) \leq 0$ as p approaches K. Consider any $v \in V_{K'}$. Then

$v(p) + u(p) \leq 1$ outside of K, for this inequality holds when p tends to ∞ and also when p approaches K. But if $u_{K'}$ does not exist, v can be chosen so that $v(p)$ is arbitrarily close to 1. It follows that $u(p) \leq 0$, so that the maximum principle is valid.

If K and K' are arbitrary, we choose K'' so that $K \cap K' \subset \text{int } K''$. Let u be as before. We have just seen that the maximum principle is valid in $W - K''$. Hence $u \leq \max\limits_{\partial K''} u$ outside of K''. If $\max\limits_{\partial K''} u$ were >0, the inequality $u \leq \max\limits_{\partial K''} u$ would also hold in $K'' - K$. Thus u would attain its maximum in $W - K$ on $\partial K''$, and necessarily at an interior point. Since this is impossible, we have $u \leq 0$ on $\partial K''$. On applying the maximum principle separately to $K'' - K$ and $W - K''$ it follows that $u \leq 0$ in $W - K$, which is what we wanted to show.

It remains to prove properties (AII) and $(AIII)$ of $g(p,p_0)$. We again use a standard local variable z with $z(p_0) = 0$. Let $m(r)$ denote the maximum of $g(p,p_0)$ when $|z(p)| = r$. From (10-2) we conclude that $m(r) + \log r$ is an increasing function of r. Hence $g(p,p_0) + \log|z(p)|$ is bounded above near p_0. On the other hand, define $v(p)$ as $-\log|z(p)| + \log r_0$ for $|z(p)| < r_0$ and zero elsewhere. It is evident that this function belongs to V_{p_0}, so that $g(p,p_0) \geq -\log|z(p)| + \log r_0$. Classically, an isolated singularity of a bounded harmonic function is removable. Property $(AIII)$ has been proved.

Denote inf $g(p,p_0)$ by c. Since we now know that $g(p,p_0) + \log|z(p)|$ has a finite limit when $p \to p_0$, we may conclude that $(1 - \epsilon)v(p) \leq g(p,p_0) - c$ for every $v \in V_{p_0}$. It follows that $c \leq 0$ and therefore in fact $c = 0$ as asserted in (AII).

Definition 10-1   An open Riemann surface with one, and hence all the properties listed in Theorem 10-2, is said to be hyperbolic. An open Riemann surface which does not have these properties is called parabolic.

For instance, a disk is hyperbolic and the whole complex plane is parabolic. The Riemann sphere is compact, and hence neither hyperbolic nor parabolic. Parabolic surfaces share many properties with compact surfaces. As an exercise, let us prove the following:

Proposition   A positive harmonic function on a parabolic surface is a constant.

PROOF Let u be positive and harmonic on a parabolic Riemann surface W. Let p and q be points on W. The function $-u$ is harmonic and bounded above. We apply the maximum principle to $-u$ on $W - \{p\}$ and $W - \{q\}$, respectively, to obtain $-u(q) \leq -u(p) \leq -u(q)$, showing that u is constant.

10-4 PROOF OF THE UNIFORMIZATION THEOREM (PART I)

Topologically equivalent surfaces may carry different conformal structures. The most obvious example is a disk and the plane which have the same topological structure but are not conformally equivalent. As a rule a topological surface can carry very many conformal structures. The uniformization theorem points out the exceptions. It asserts that a topological sphere has only one conformal structure and that a topological disk has two. It is convenient to combine these assertions into a single statement concerning simply connected surfaces.

Theorem 10-3 (**The uniformization theorem**) Every simply connected Riemann surface is conformally equivalent to a disk, to the complex plane, or to the Riemann sphere.

Since the existence of Green's function is a conformally invariant property, it is clear from the beginning that a Riemann surface can be conformally equivalent to a disk only if it is hyperbolic and to the whole plane only if it is parabolic. The spherical type is characterized by compactness. The three cases will be considered separately.

The hyperbolic case. W is a simply connected Riemann surface whose Green's function $g(p,p_0)$ exists for every p_0. Each $p \neq p_0$ has a neighborhood U_α which does not contain p_0 and is conformally equivalent to a disk. Let h_α be a conjugate harmonic function of $g(p,p_0)$ in U_α; it is determined up to an additive constant. The function $f_\alpha(p) = e^{-(g+ih_\alpha)}$ is analytic in U_α and determined up to a constant factor of absolute value 1.

In a neighborhood U_{α_0} of p_0 we can similarly determine a conjugate harmonic function $h_{\alpha_0}(p)$ of $g(p,p_0) + \log |z(p)|$ for some choice of the local variable z. We write $f_{\alpha_0}(p) = e^{-(g+ih_{\alpha_0})}$ for $p \neq p_0$ and $f_{\alpha_0}(p_0) = 0$.

We are now in a position to apply Theorem 9-5. Indeed, the U_α form an open covering of W, and in each U_α we have defined a family of functions f_α. In the overlap of U_α and U_β the quotient f_α/f_β has constant absolute value and is therefore itself a constant on each component of

$U_\alpha \cap U_\beta$. It follows that f_α and f_β are either everywhere equal or everywhere unequal on each component of $U_\alpha \cap U_\beta$, and if f_α is given in a component the constant in f_β can be adjusted so that $f_\alpha = f_\beta$. The theorem guarantees the existence of an analytic function $f(p,p_0)$ on W, which vanishes for $p = p_0$ and satisfies $\log |f(p,p_0)| = -g(p,p_0)$ for $p \neq p_0$. Observe that $|f(p,p_0)| < 1$ and that $f(p,p_0) = 0$ only for $p = p_0$.

To prove the theorem it suffices to show that $f(p,p_0)$ is one to one, for then W is conformally equivalent to a bounded plane region and we can appeal to the Riemann mapping theorem to conclude that W is conformally equivalent to the unit disk. Actually, the standard proof of the mapping theorem would show that $f(p,p_0)$ is itself a mapping onto the unit disk.

The idea for the remaining part of the proof is to compare $f(p,p_0)$ with $f(p,p_1)$ for $p_1 \neq p_0$. Consider the function

$$F(p) = [f(p,p_0) - f(p_1,p_0)] : [1 - \bar{f}(p_1,p_0)f(p,p_0)]. \qquad (10\text{-}3)$$

Because $|f(p_1,p_0)| < 1$, the fraction on the right is pole-free, and F is analytic on W with $|F| < 1$ and $F(p_1) = 0$.

We recall from Sec. 10-1 that every $v \in V_{p'}$ is subharmonic, vanishes near the ideal boundary, and satisfies $\overline{\lim_{p \to p_1}} [v(p) + \log |z_1(p)|] < \infty$, where z_1 is a local variable with $z_1(p_1) = 0$. Since $F(p)/z_1(p)$ is regular at p_1, it follows that $\overline{\lim_{p \to p_1}} [v(p) + (1 + \epsilon) \log |F(p)|] = -\infty$ for $\epsilon > 0$. By use of the maximum principle we conclude that $v(p) + (1 + \epsilon) \log |F(p)| \leq 0$ on W. On passing to the limit we obtain

$$g(p,p_1) + \log |F(p)| \leq 0. \qquad (10\text{-}4)$$

This can also be written as $|F(p)| \leq |f(p,p_1)|$. For $p = p_0$ the inequality yields $|f(p_1,p_0)| \leq |f(p_0,p_1)|$. But p_0 and p_1 are interchangeable, so we have in fact proved that $|f(p_1,p_0)| = |f(p_0,p_1)|$.

As a result (10-4) becomes an equality for $p = p_0$. The left-hand member is thus a harmonic function which attains its maximum. We conclude that it must be identically zero so that $|F(p)| = |f(p,p_1)|$, and hence $F(p) = e^{i\theta}f(p,p_1)$ with constant real θ. We further conclude that $F(p) = 0$ only for $p = p_1$, and by (10-3) this means that $f(p,p_0) = f(p_1,p_0)$ only for $p = p_1$. We have shown that $f(p,p_0)$ is indeed univalent.

The parabolic case.   The difficulty with this case is that we do not know a priori the existence of a single nonconstant harmonic or subharmonic function even if we allow for a singularity. As a substitute for the Green's function we shall need a function $u(p,p_0)$ which is harmonic for $p \neq p_0$ and behaves like Re $1/z(p)$ in terms of a local variable with $z(p_0) = 0$. It would be tempting to construct $u(p,p_0)$ by means of a Perron family which might be defined by a condition $\overline{\lim_{p \to p_0}} [v(p) - \text{Re } 1/z(p)] \leq 0$. This approach must be abandoned because there is no easy way of ascertaining that the family in question is not empty. Instead we have to rely on a less direct method which in its essential features goes back to Neumann [40].

Lemma 10-1 Suppose that $u(z)$ is harmonic for $\rho \leq |z| \leq 1$ and constant on $|z| = \rho$. Let $S_r(u) = \max\limits_{|z|=r} u(z) - \min\limits_{|z|=r} u(z)$ be the oscillation of u on $|z| = r$. Then

$$S_r(u) \leq q(r)S_1(u), \tag{10-5}$$

where $q(r)$ depends only on r and $q(r) \to 0$ when $r \to 0$.

PROOF   We may assume that the maximum and minimum of u on $|z| = r$ are attained at conjugate points z_0 and $\bar{z}_0$. Consider the function $u(z) - u(\bar{z})$ which is harmonic in the upper half annulus $\rho \leq |z| \leq 1$, Im $z \geq 0$. It is zero on the real axis and the inner half circle, and it is $\leq S_1(u)$ on the outer half circle. At z_0 it is equal to $S_r(u)$. It is majorized by the harmonic function in the full half disk which is $S_1(u)$ on $|z| = 1$ and zero on the diameter. In terms of the angle α indicated in Fig. 10-1 we thus have

$$S_r(u) \leq \frac{2}{\pi} (\pi - \alpha)S_1(u).$$

Here α assumes its minimum for fixed r when $z_0 = ir$, and we obtain

$$S_r(u) \leq \left(\frac{4}{\pi} \text{arc tan } r\right) S_1(u),$$

which is of the desired form.

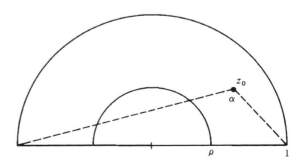

FIGURE 10-1

Lemma 10-2 Let $z(p)$ be a local variable with $z(p_0) = 0$ and denote the inverse image of $|z| < \rho$ by Δ_ρ. If W is parabolic, there exists a unique bounded harmonic function u_ρ on $W - \bar{\Delta}_\rho$ with the boundary values Re $1/z(p)$.

PROOF   Perron's method is applicable to the family of bounded subharmonic functions v on $W - \bar{\Delta}_\rho$ that satisfy $v \leq$ Re $1/z(p)$ on the boundary of Δ_ρ. These functions are uniformly bounded because the maximum principle is valid in $W - \bar{\Delta}_\rho$, and this is sufficient for $u_\rho = \sup v$ to be harmonic. A well-known elementary argument (see C.A., p. 241) shows that u_ρ has the right boundary values. The uniqueness is a consequence of the maximum principle.

Lemma 10-3 The function u_ρ of the preceding lemma satisfies

$$\int_0^{2\pi} \frac{\partial}{\partial r} u_\rho(re^{i\theta}) \, d\theta = 0. \tag{10-6}$$

PROOF   Let $D \subset W$ be a relatively compact region with smooth boundary such that $\bar{\Delta}_\rho \subset D$. Denote the harmonic measure of $\partial\Delta_\rho$ with respect to $D - \bar{\Delta}_\rho$ by ω. Then

$$\int_{\partial\Delta_\rho} \left(\omega \frac{\partial u_\rho}{\partial n} - u_\rho \frac{\partial \omega}{\partial n} \right) ds = \int_{\partial D} \left(\omega \frac{\partial u_\rho}{\partial n} - u_\rho \frac{\partial \omega}{\partial n} \right) ds, \tag{10-7}$$

where the line integrals and normal derivatives can be expressed in terms of local variables.

We know that $|u| \leq 1/\rho$ by the maximum principle. Note further that $\omega = 1$ on $\partial\Delta_\rho$, $\omega = 0$ on ∂D, and that $\partial\omega/\partial n$ has constant sign on $\partial\Delta_\rho$ and on ∂D. With these observations (10-7) leads to

$$\left| \int_{\partial\Delta_\rho} \frac{\partial u_\rho}{\partial n} \, ds \right| \leq \frac{1}{\rho} \left| \int_{\partial\Delta_\rho} \frac{\partial \omega}{\partial n} \, ds \right| + \frac{1}{\rho} \left| \int_{\partial D} \frac{\partial \omega}{\partial n} \, ds \right| = \frac{2}{\rho} \left| \int_{\partial\Delta_\rho} \frac{\partial \omega}{\partial n} \, ds \right|.$$

Now we let D expand. Because $\bar{\Delta}_\rho$ has no harmonic measure, ω will tend to 1, uniformly in a neighborhood of $\partial\Delta_\rho$ (note that ω can be extended by the symmetry principle). Hence $\partial\omega/\partial n$ converges uniformly to zero on $\partial\Delta_\rho$, and we conclude that

$$\int_{\partial\Delta_\rho} \frac{\partial u_\rho}{\partial n} \, ds = 0.$$

In terms of the local variable z this is (10-6) for $r = \rho$. However, the integral in (10-6) is independent of r.

Lemma 10-4 As $\rho \to 0$ the functions u_ρ tend to a harmonic function u on $W - \{p_0\}$ which is bounded outside of every Δ_ρ and satis-

fies $\lim_{p \to p_0} [u(p) - \operatorname{Re} 1/z(p)] = 0$. The function u is uniquely determined by these conditions.

PROOF We assume that the range of $z(p)$ contains $|z| \leq 1$. Apply
Lemma 10-1 to $u - \operatorname{Re} 1/z$. We obtain

$$S_r \left(u - \operatorname{Re} \frac{1}{z} \right) \leq q(r) S_1 \left(u - \operatorname{Re} \frac{1}{z} \right), \tag{10-8}$$

and hence

$$S_r(u_\rho) - \frac{2}{r} \leq q(r)[S_1(u_\rho) + 2]. \tag{10-9}$$

On the other hand, the maximum principle on $W - \bar{\Delta}_\rho$ yields

$$S_1(u_\rho) \leq S_r(u_\rho). \tag{10-10}$$

We conclude from (10-9) and (10-10) that

$$S_1(u_\rho) \leq \frac{2q(r) + 1}{1 - q(r)}. \tag{10-11}$$

For a fixed $r = r_0 < 1$ it follows from (10-11) that $S_1(u_\rho)$ is less than
a constant C independent of ρ. We return to (10-8) to obtain

$$S_r u_\rho - \operatorname{Re} \frac{1}{z} \leq (C + 2)q(r). \tag{10-12}$$

Lemma 10-3 shows that the mean value of u_ρ over a circle $|z| = r$ is inde-
pendent of r. Since $\operatorname{Re} 1/z$ has mean value zero, it follows that the mean of
$u_\rho - \operatorname{Re} 1/z$ is also zero, and we conclude from (10-12) that

$$\max_{|z| = r} \left| u_\rho - \operatorname{Re} \frac{1}{z} \right| \leq (C + 2)q(r), \tag{10-13}$$

and hence also

$$\max_{|z| = r} |u_\rho - u_{\rho'}| \leq 2(C + 2)q(r) \tag{10-14}$$

for ρ, $\rho' < r$. By the maximum principle $|u_\rho - u_{\rho'}|$ has the same bound
in the whole complement of Δ_r. This proves the existence of $u = \lim_{\rho \to 0} u_\rho$
as a uniform limit outside of any neighborhood of p_0.

From (4-13) and (4-14) we now obtain

$$\max_{|z| = r} \left| u - \operatorname{Re} \frac{1}{z} \right| \leq (C + 2)q(r)$$

and

$$\max_{|z| = r} |u_\rho - u| \leq 2(C + 2)q(r).$$

The first inequality proves that $u - \text{Re } 1/z \to 0$ for $p \to p_0$, and the second shows that u is bounded outside of Δ_ρ. Finally, the uniqueness of u follows by the maximum principle.

10-5 PROOF OF THE UNIFORMIZATION THEOREM (PART II)

We continue our discussion of the parabolic case. Having constructed our function u we observe that every point on W, including p_0, has a neighborhood in which u has a conjugate function v which is determined up to an additive constant. By use of Theorem 9-5 we can form a global meromorphic function $f = u + iv$ with a development

$$f(p) = \frac{1}{z} + az + \cdots \tag{10-15}$$

in terms of the local variable at p_0. This normalization determines f uniquely.

Suppose that we replace z by $\tilde{z} = -iz$. There is a corresponding function $\tilde{f}$, and in terms of the original variable we have

$$\tilde{f}(p) = \frac{i}{z} + bz + \cdots . \tag{10-16}$$

We shall show that $\tilde{f} = if$. For this purpose we consider an arbitrary Δ_ρ and assume that $|\text{Re } f| \leq M$ and $|\text{Re } \tilde{f}| \leq M$ outside of Δ_ρ. There exists a point $p_1 \neq p_0$ in Δ_ρ such that $\text{Re } f(p_1) > M$ and $\text{Re } \tilde{f}(p_1) > M$; it suffices to choose p_1 close to p_0 with $\arg z(p_1) = \pi/4$. Then $f(p) \neq f(p_1)$ for p outside of Δ_ρ, and since $\text{Re } [f(p) - f(p_1)] < 0$ on $\partial \Delta_\rho$, it follows by the argument principle that $f(p) - f(p_1)$ has p_1 as its only zero, this zero being simple. The same applies to $\tilde{f}$. We therefore have expansions of the form

$$F(p) = \frac{f(p)}{f(p) - f(p_1)} = \frac{A}{z - z_1} + B + \cdots$$

$$F(p) = \frac{\tilde{f}(p)}{\tilde{f}(p) - \tilde{f}(p_1)} = \frac{\tilde{A}}{z - z_1} + \tilde{B} + \cdots . \tag{10-17}$$

Due to our choice of p_1 we have

$$|F(p)| \leq 1 + \frac{|f(p_1)|}{\text{Re } f(p_1) - M}$$

for p outside of Δ_ρ, and a similar bound for $\tilde{F}$. Hence the linear combination $\tilde{A}F - A\tilde{F}$ is bounded and analytic on the whole surface. Since W is parabolic, this function must reduce to a constant, and we conclude that $\tilde{f} = T(f)$, where T is a fractional linear transformation. The developments (10-15) and (10-16) show that the only possibility is to have $\tilde{f} = if$.

The relation $\tilde{f} = if$ shows that f itself, and not only Re f, is bounded outside of Δ_ρ. Suppose that $|f(p)| < M_1$ outside of Δ_ρ and choose any $p_1 \neq p_0$ in Δ_ρ such that $|f(p_1)| > M_1$. Then F, as defined by the first line in (10-17), is again bounded outside of Δ_ρ, and the argument principle (or Rouché's theorem) shows that the only singularity of F is a simple pole at p_1.

Although f is not uniquely determined by p_0, we shall make a definite choice of the local variable z and denote the corresponding f by $f(p,p_0)$. For $p_1 \in \Delta_\rho$ we then define $f(p,p_1)$ by use of the same z. We compare the developments of $f(p,p_1)$ and $F(p)$ which have the same singular point. Because the surface is parabolic, it follows that $F(p) = af(p,p_1) + b$ with constant coefficients, and hence $f(p,p_1)$ is a linear fractional transformation of $f(p,p_0)$. This is true even when p_1 is not close to p_0, for we can pass from p_0 to p_1 through a sequence of intermediate points, each close to the preceding one. Let us write, explicitly, $f(p,p_1) = S[f(p,p_0)]$.

It is now easy to conclude that $f(p,p_0)$ is one to one. In fact, suppose that $f(p,p_0) = f(p_1,p_0)$. With S defined as above we then have

$$f(p,p_1) = S[f(p,p_0)] = S[f(p_1,p_0)] = f(p_1,p_1) = \infty,$$

and hence $p = p_1$ since p_1 is the only pole of $f(p,p_1)$.

We have shown that W is conformally equivalent to an open subset of $\mathbf{C} \cup \{\infty\}$. This set cannot be the whole Riemann sphere, for then W would be compact. Neither can its complement consist of more than one point, for if it did the Riemann mapping theorem would show that W is hyperbolic. An inversion throws the complementary point to ∞, and we have completed our task of mapping W on the whole complex plane.

The compact case. One possibility would be to remove a point p_0 and show that $W - \{p_0\}$ is parabolic. The objection is that it is not quite trivial to prove by topological methods that $W - \{p_0\}$ is simply connected.

The alternative is to repeat the proof given for the parabolic case. Lemmas 10-2 to 10-4 remain in force, the proofs being somewhat simpler because only the classic maximum principle is needed. The function $f(p,p_0)$ is constructed as before, and the same reasoning shows that it is a one-to-one mapping. The range of $f(p,p_0)$ is an open and compact subset of the Riemann sphere, and hence the whole sphere.

REMARK   In our definition of surface and Riemann surface we never required that the underlying space satisfy the second axiom of countability, nor did our proof overtly or covertly make use of this property. It is indeed a remarkable feature of the Perron method that it uses only local constructions so that global countability never enters the picture. In the proof of Lemma 10-3 we used an exhaustion by expanding regions D, but D expands in the sense of partial ordering by inclusion and no sequences are required. Therefore, by proving the uniformization theorem we have shown that every simply connected Riemann surface satisfies the second countability axiom, and by passing to the universal covering it follows that the same is true for an arbitrary Riemann surface. This observation is due to Radó [55].

10-6   ARBITRARY RIEMANN SURFACES

We shall now drop the condition of simple connectivity. We showed in Chap. 9 that every Riemann surface W has an essentially unique universal covering surface $\tilde{W}$. It was defined by the condition $\pi_1(\tilde{W}) = 1$, and this property characterizes $\tilde{W}$ up to conformal equivalence as the only simply connected covering surface of W.

Because $\tilde{W}$ is simply connected, we can apply the uniformization theorem to conclude that $\tilde{W}$ is conformally equivalent to either the Riemann sphere, the complex plane, or the unit disk. Since conformal mapping does not change the relevant properties of a Riemann surface, we are free to assume that $\tilde{W}$ *is* one of these three surfaces. For the moment we wish to give a parallel treatment of all cases, and for this reason we shall not yet specify whether $\tilde{W}$ is hyperbolic, parabolic, or compact. In any case points on $\tilde{W}$ can be regarded as complex numbers z, possibly including $z = \infty$.

More precisely, the universal covering surface is a pair $(\tilde{W}, f)$, where $f: \tilde{W} \to W$ is the projection map. We can regard $f(z)$ as an analytic function on $\tilde{W}$ with values on W.

Recall that a homeomorphism $\varphi: \tilde{W} \to W$ is a cover transformation if $f \circ \varphi = f$. We remarked at the beginning of Sec. 9-5 that every cover transformation is a conformal homeomorphism, and according to Theorem 9-3 a cover transformation has no fixed points (unless it is the identity).

In all three cases, namely, for the sphere, the plane, and the disk, we know that all conformal self-mappings are given by linear transformations $\varphi(z) = (az + b)/(cz + d)$, $ad - bc \neq 0$. Every such mapping has at least one fixed point on the sphere. Therefore, if $\tilde{W}$ is the sphere, φ can only be the identity. If $\tilde{W}$ is the plane, the only fixed point must be at ∞, and this implies $\varphi(z) = z + b$ so that φ is a parallel translation. Finally, if $\tilde{W}$ is the unit disk, the fixed points must lie on the unit circle.

In this case φ is either a parabolic or a hyperbolic transformation of the form $\varphi(z) = (az + b)/(\bar{b}z + \bar{a})$. As such φ may be regarded as a non-euclidean motion which is not a rotation.

By Theorem 9-4 the cover transformations form a group which in the present case $(D = 1)$ is isomorphic to $\pi_1(W)$. If $\tilde{W}$ is the sphere, we conclude that $\pi_1(W) = 1$. Since $\tilde{W}$ is compact, so is its projection W, and we conclude by the uniformization theorem that W is conformally a sphere, in one-to-one correspondence with $\tilde{W}$. This trivial case can be ignored. In the remaining cases $\pi_1(W)$ can be represented as a group of parallel translations of the euclidean plane, or as a group of fixed point-free motions of the noneuclidean plane.

There is an additional property of the cover transformations which we have not yet taken into account. Each point $p \in \tilde{W}$ has a neighborhood $\tilde{V}$ which is in one-to-one correspondence with its projection $f(\tilde{V})$. If φ is a cover transformation other than the identity, it follows that $\tilde{V}$ and $\varphi(\tilde{V})$ are disjoint. Indeed, if $p \in \tilde{V} \cap \varphi(\tilde{V})$, we would have $p = \varphi(q)$ with $p, q \in \tilde{V}$, and hence $f(p) = f[\varphi(q)] = f(q)$, which is possible only if $p = q$. But then $p = \varphi(p)$ is a fixed point, so that φ must be the identity. To repeat, every point of $\tilde{W}$ has a neighborhood which does not meet its images under the cover transformation. We express this by saying that the group of cover transformations is properly discontinuous on $\tilde{W}$.

When $\tilde{W}$ is the plane, we know now that the group Γ of cover transformations is a properly discontinuous group of parallel translations. It is classic that there are only three types of such groups: (1) the identity, (2) the infinite cyclic group generated by $\varphi(z) = z + b$, $b \neq 0$, (3) the abelian group generated by $\varphi_1(z) = z + b_1$ and $\varphi_2(z) = z + b_2$ with a non-real ratio b_2/b_1. The surface W is recovered by identifying points that correspond to each other under the transformations in Γ. In case (1) W is the plane; in case (2) it is an infinite cylinder, conformally equivalent with the punctured plane; and in case (3) it is a torus obtained by identifying opposite sides of a parallelogram. The theory of analytic functions on the torus is equivalent to the theory of elliptic functions.

In all cases except the ones listed above $\tilde{W}$ is the disk. The group Γ is a properly discontinuous group of fixed point-free linear transformations which map the disk on itself. Conversely, if Γ is such a group, we obtain a Riemann surface by identifying points that are equivalent under the group. We collect all this in a theorem.

Theorem 10-4 If a Riemann surface W is not conformally equivalent to a sphere, a plane, or a punctured plane, there exists a properly discontinuous group Γ of fixed point-free linear transformations mapping the unit disk Δ onto itself such that the Riemann surface Δ/Γ

obtained by identifying equivalent points under the group is conformally equivalent to W.

The theory of analytic functions on W becomes the theory of automorphic functions under the group Γ. The hyperbolic metric of the disk carries over to the Poincaré metric on W with constant curvature -1. In particular, we conclude that every plane region whose complement with respect to the plane has more than one point carries a Poincaré metric (see I-1-7).

NOTES   All classic proofs of the uniformization theorem make use of the "oil speck" method, which consists in exhausting the surface by a sequence of relatively compact subregions. The Perron method makes it possible to dispense with this method and to define Green's function and harmonic measure directly. Otherwise our proof is akin to the alternating method of Schwarz and Neumann, for the comparisons in Theorem 10-2 use the same classic estimates. In Secs. 10-4 and 10-5 the reasoning is modeled on Heins [31].

BIBLIOGRAPHY

1   AHLFORS, L. V.: An Extension of Schwarz' Lemma, *Trans. Am. Math. Soc.*, **43**, 359–364 (1938).

2   AHLFORS, L. V.: Untersuchungen zur Theorie der konformen Abbildung und der ganzen Funktionen, *Acta Soc. Sci. Fenn., Nov. Ser. A1*, **9**, 1–40 (1930).

3   AHLFORS, L. V., and BEURLING, A.: Invariants conformes et problèmes extrémaux, *10th Scand. Congr. Math.*, 341–351 (1946).

4   AHLFORS, L. V., and BEURLING, A.: Conformal Invariants and Function-Theoretic Nullsets, *Acta Math.*, **83**, 101–129 (1950).

5   AHLFORS, L. V., and SARIO, L. "Riemann Surfaces," Princeton University Press, 1960.

6   BEURLING, A.: Études sur un problème de majoration, Thèse, Uppsala, 1933.

7   BIEBERBACH, L.: Über die Koeffizienten derjenigen Potenzreihen, welche eine schlichte Abbildung des Einheitskreises vermitteln, *Sitz. Ber. Preuss. Akad. Wiss.*, **138**, 940–955 (1916).

8   BLOCH, A.: Les théorèmes de M. Valiron sur les fonctions entières et la théorie de l'uniformisation, *Ann. Fac. Sci. Univ. Toulouse*, **III**, 17 (1925).

9 BRELOT, M.: La théorie moderne du potential, *Ann. Inst. Fourier*, **4**, 113–140 (1952).

10 CARATHÉODORY, C.: Untersuchungen über die konformen Abbildungen von festen und veränderlichen Gebieten, *Math. Ann.*, **52**:(1), 107–144 (1912).

11 CARATHÉODORY, C.: Über die Winkelderivierten von beschränkten analytischen Funktionen, *Sitz. Ber. Preuss. Akad., Phys.-Math.*, **IV**, 1–18 (1929).

12 CARLEMAN, T.: Sur les fonctions inverses des fonctions entières, *Ark. Mat. Astr. Fys.*, **15**, 10 (1921).

13 CARLESON, L.: An Interpolation Problem for Bounded Analytic Functions, *Am. J. Math.*, **80**:(4), 921–930 (1958).

14 CHARZYNSKI, Z., and SCHIFFER, M.: A New Proof of the Bieberbach Conjecture for the Fourth Coefficient, *Arch. Rat. Mech. Anal.*, **5**, 187–193 (1960).

15 COURANT, R.: "Dirichlet's Principle, Conformal Mapping, and Minimal" Surfaces. With an appendix by M. Schiffer. Interscience, New York, 1950.

16 CURTISS, J.: Faber Polynomials and the Faber Series, *Am. Math. Mon.*, **78**:(6), 577–596 (1971).

17 DENJOY, A.: Sur une classe de fonctions analytiques, *C. R.* **188,** 140, 1084 (1929).

18 DOUGLAS, J.: Solution of the Problem of Plateau, *Trans. Am. Math. Soc.*, **33**, 263–321 (1931).

19 FEKETE, M.: Über den transfiniten Durchmesser ebener Punktmengen I–III, *Math. Z.*, **32**, 108–114, 215–221 (1930); **37**, 635–646 (1933).

20 FROSTMAN, O.: Potentiel d'équilibre et capacité des ensembles, *Medd. Lunds Mat. Sem.*, **3**, 1–115 (1935).

21 GARABEDIAN, P., and SCHIFFER, M.: A Proof of the Bieberbach Conjecture for the Fourth Coefficient, *J. Rat. Mech. Anal.*, **4**, 427–465 (1955).

22 GOLUSIN, G. M.: "Geometrische Funktionentheorie," Deutscher Verlag, Berlin, 1957.

23 GRONWALL, T. H.: Some Remarks on Conformal Representation, *Ann. Math.*, **16**, 72–76 (1914–1915).

24 GRÖTZSCH, H.: Eleven papers in *Ber. Verh. Sächs. Akad. Wiss. Leipzig, Math. Phys.* (1928–1932).

25 GRUNSKY, H.: Koeffizientenbedingungen für schlicht abbildende meromorphe Funktionen, *Math. Z.*, **45**, 29–61 (1939).

26 HARDY, G. H., and LITTLEWOOD, J. E.: A Maximal Theorem with Function-Theoretic Applications, *Acta Math.*, **54**, 81–116 (1930).

27 HAYMAN, W. K.: "Multivalent Functions," Cambridge Tracts, 48, Cambridge University Press, 1958.

28 HEINS, M.: "Selected Topics in the Classical Theory of Functions of a Complex Variable," Holt, Rinehart and Winston, New York, 1962.

29 HEINS, M.: On a Problem of Walsh concerning the Hadamard Three Circles Theorem, *Trans. Am. Math. Soc.*, **55**:(1), 349–372 (1944).

30 HEINS, M.: The Problem of Milloux for Functions Analytic throughout the Interior of the Unit Circle, *Am. J. Math.*, **57**:(2), 212–234 (1945).

31 HEINS, M.: The Conformal Mapping of Simply Connected Riemann Surfaces, *Ann. Math.*, **50**, 686–690 (1949).

32 JENKINS, J.: On Explicit Bounds in Schottky's Theorem, *Can. J. Math.*, **7**, 80–99 (1955).

33 JENKINS, J.: Some Area Theorems and a Special Coefficient Theorem, *Ill. J. Math.*, **8**, 80–99 (1964).

34 LANDAU, E.: Der Picard-Schottkysche Satz und die Blochsche Konstante, *Sitz. Ber. Preuss. Akad., Phys.-Math.* (1926).

35 LANDAU, E., and VALIRON, G.: A Deduction from Schwarz's Lemma, *J. London Math. Soc.*, **4.3**, 162–163 (1929).

36 LÖWNER, K.: Untersuchungen über schlichte konforme Abbildungen des Einheitskreises I, *Math. Ann.*, **89**, 103–121 (1923).

37 MARTY, F.: Sur les modules des coefficients de MacLaurin d'une fonction univalente, C. R. **198**, 1569–1571 (1934).

38 MILLOUX, H.: Sur le théorème de Picard, *Bull. Soc. Math. Fr.*, **B53**, 181–207 (1925).

39 MYRBERG, P. J.: Über die Existenz der Greenschen Funktionen auf einer gegebenen Riemannschen Fläche, *Acta Math.*, **61**, 39–79 (1933).

40 NEUMANN, C.: "Theorie der Abelschen Integrale," Teubner, Leipzig, 1884.

41 NEVANLINNA, F., and NEVANLINNA, R.: Über die Eigenschaften einer analytischen Funktion in der Umgebung einer singulären Stelle oder Linie, *Acta Soc. Sci. Fenn.*, **50**, 5 (1922).

42 NEVANLINNA, R.: Über beschränkte Funktionen die in gegebenen Punkten vorgeschriebene Werte annehmen, *Ann. Acad. Sci. Fenn.*, **13**, No. 1 (1919).

43 NEVANLINNA, R.: Über beschränkte analytische Funktionen, *Ann. Acad. Sci. Fenn.*, **32**, No. 7, 1–75 (1929).

44 NEVANLINNA, R.: Über eine Minimumaufgabe in der Theorie der konformen Abbildung, *Göttinger Nachr.*, **I.37**, 103–115 (1933).

45 NEVANLINNA, R.: Das harmonische Mass von Punktmengen und seine Anwendung in der Funktionentheorie, *8th Scand. Math. Congr.*, *Stockholm* (1934).

46 NEVANLINNA, R.: "Eindeutige analytische Funktionen," Springer, Berlin, 1936.

47 NEVANLINNA, R.: Über die schlichten Abbildungen des Einheitskreises, *Övers. Finska Vetensk.-Soc. Förh.*, **62A,** 6 (1919–1920).

48 OHTSUKA, M.: "Dirichlet Problem, Extremal Length and Prime Ends," Van Nostrand, New York, 1970.

49 OSTROWSKI, A.: Über allgemeine Konvergenzsätze der komplexen Funktionentheorie, *Jahresber. Deutsche Math.-Ver.*, **32,** 185–194 (1923).

50 PICK, G.: Über eine Eigenschaft der konformen Abbildung kreisförmiger Bereiche, *Math. Ann.*, **77,** 1–6 (1915).

51 PICK, G.: Über die Beschränkungen analytischer Funktionen, welche durch vorgeschriebene Werte bewirkt werden, *Math. Ann.*, **77,** 7–23 (1915).

52 POLYA, G., and SZEGÖ, G.: Über den transfiniten Durchmesser (Kapazitätskonstante) von ebenen und räumlichen Punktmengen, *J. Reine Angew. Math.*, **165,** 4–49 (1931).

53 POMMERENKE, CH.: Über die Faberschen Polynome schlichter Funktionen, *Math. Z.*, **85,** 197–208 (1964).

54 POMMERENKE, CH.: On the Grunsky Inequalities for Univalent Functions, *Arch. Rat. Mech. Anal.*, **35:**(3), 234–244 (1969).

55 RADÓ, T.: Über den Begriff der Riemannschen Fläche, *Acta Szeged*, **2,** 101–121 (1925).

56 SCHAEFFER, A. C., and SPENCER, D. C.: Coefficient Regions for Schlicht Functions, *A.M.C. Colloq. Publ.*, **35** (1950).

57 SCHIFFER, M.: A Method of Variation within the Family of Simple Functions, *Proc. London Math. Soc.*, **2:**(44), 450–452 (1938).

58 SCHLESINGER, E.: Conformal Invariants and Prime Ends, *Am. J. Math.*, **80,** 83–102 (1958).

59 SCHWARZ, H. A.: "Gesammelte Abhandlungen," Vol. II, Springer, Berlin, 1890.

60 SPRINGER, G.: "Introduction to Riemann Surfaces," Addison-Wesley, Reading, Mass., 1957.

61 SZEGÖ, G.: Bemerkungen zu einer Arbeit von Herrn M. Fekete, *Math. Z.*, **21,** 203–208 (1924).

62 TEICHMÜLLER, O.: Eine Verschärfung des Dreikreisesatzes, *Deutsche Math.*, **4:**(1), 16–22 (1939).

63 TEICHMÜLLER, O.: Extremale quasikonforme Abbildungen und quadratische Differentiale, *Abh. Preuss. Akad. Wiss., Math.-Nat.*, **22,** 1–197 (1939).

64 TEICHMÜLLER, O.: Über Extremalprobleme der konformen Geometrie, *Deutsche Math.*, **6,** 50–77 (1941).

65 TEICHMÜLLER, O.: Untersuchungen über konforme und quasikonforme Abbildungen, *Deutsche Math.*, **3,** 621–678 (1938).

66 WEYL, H.: "Die Idee der Riemannschen Fläche," 1st ed., Teubner, Berlin, 1913; 2d ed., 1923; 3d ed., Stuttgart, 1955.

Index

$|a_3| \leq 3$, 96
$|a_4| \leq 4$, 87
Ahlfors, L. V., 21, 25, 76, 81, 135, 138, 152
Angular limit, 9
Annulus, 71
Area theorem, 82

Beurling, A., 36, 42, 43, 48, 61, 81, 152
Bieberbach, L., 83, 152
Bloch, A., 152
Bloch's constant, 14
Bloch's theorem, 14, 22
Brelot, M., 36, 153

Capacity, 23, 25, 27
Carathéodory, C., 7, 11, 21, 57, 153
Carleman, T., 42, 48, 153
Carleson, L., 5, 153
Cartan, H., 36
Charzynski, Z., 87, 153
Chebyshev polynomial, 24
Comparison principle, 53
Complex manifold, 125
Composition laws, 54
Configuration, 70
Conformal invariant, 2
    relative, 28
Conformal structure, 126
Convex region, 5
Courant, R., 106, 153
Cover transformation, 132
Covering surface, 125, 130
Curvature, 12

Denjoy, A., 5, 153
Dirichlet integral, 32, 79
Dirichlet-Neumann problem, 65, 78
Distortion theorem, 84
Douglas, J., 36, 153

Energy integral, 25
Equimeasurable, 31
Extremal distance, 52, 65
    conjugate, 53
    reduced, 78
Extremal function, 107
Extremal length, 50
Extremal metric, 61

Faber polynomials, 83, 90
Fekete, M., 36, 153
Frostman, O., 36, 153
Fundamental group, 128, 129

Γ structure, 114, 119
Garabedian, P., 87, 153
Gauss-Bonnet formula, 112
Geodesic polygon, 112
Golusin, G. M., 85, 87, 90, 91, 153
Green's function, 25, 137, 139
    variation of, 98
Gronwall, T. H., 83, 153
Grötzsch, H., 50, 72, 73, 81, 153
Grunsky, H., 85, 87, 90, 91, 153

Hadamard's three-circle theorem, 39, 44
Hardy, G. H., 36, 153
Harmonic measure, 37, 139
Harnack's principle, 27
Hayman, W. K., 5, 6, 36, 153
Heins, M., 21, 42, 45, 136, 151, 154
Herglotz, 11
Homotopic, 129
Homotopy class, 129
Horocycle, 7
Hyperbolic geometry, 2
Hyperbolic metric, 13
Hyperbolic Riemann surface, 141, 142

Ideal boundary, 138
Interior variation, 99

Jenkins, J., 21, 90, 154
Julia's lemma, 8

Klein, F., 135
Koebe theorem, 72, 84, 85, 136

Landau, E., 21, 154
Lavrentiev, M. A., 90
Length-area principle, 50
Lindelöf's maximum principle, 38, 40
Littlewood, J. E., 36, 153
Local variable, 126
Local parameter, 25
Löwner, K., 93, 97, 154
Löwner's differential equation, 96
Löwner's lemma, 12
Löwner's method, 93

Marty, F., 105, 154
Maximum principle, 139
Milloux, H., 154
Milloux's problem, 41
Modulus, 51, 70
Monodromy theorem, 130
Myrberg, P. J., 154

Neumann, C., 144, 151, 154
Nevanlinna, F., 48, 154
Nevanlinna, R., 4, 5, 41, 48, 90, 154
Noneuclidean distance, 2
Noneuclidean geometry, 2
Noneuclidean metric, 1

Ohtsuka, M., 81, 154
Open surface, 138
Ostrowski, A., 48, 154

Parabolic surface, 141, 143
Picard-Schottky theorem, 19
Picard theorem, 19, 21

Pick, G., 3, 21, 154
Poincaré, J. H., 21
Poincaré metric, 16, 151
Polya, G., 36, 154
Pommerenke, Ch., 21, 36, 90, 154
Potential, 24
Prime ends, 57

Quadratic differential, 110
Quadrilateral, 52, 71

Radó, T., 149, 154
Regular covering surface, 133
Riemann sphere, 127
Riemann surface, 125
Robin constant, 25, 27, 30, 35

Sario, L., 135, 152
Schaeffer, A. C., 115, 124, 154
Schiffer, M., 87, 98, 153, 154
Schiffer variation, 98, 106
Schlesinger, E., 81, 154 .
Schlicht function, 82, 84
Schwarz, H. A., 151, 154
Schwarz-Pick theorem, 3
Schwarz's lemma, 1, 3, 8, 11, 13, 21, 94
Simply connected, 130, 134
Spencer, D. C., 115, 124, 154
Springer, G., 135, 154
Subgroup, 130
Symmetrization, 31
Szegö, G., 36, 154

Teichmüller, O., 45, 48, 72–76, 154
Trajectory, 110
Transfinite diameter, 23
Two-constant theorem, 39

Ultrahyperbolic metric, 12
Uniformization theorem, 137, 142, 147
Univalent function, 82
Universal covering surface, 132

Weyl, H., 154

ERRATA

Unfortunately, a substantial number of typographical errors escaped detection when the book first went into print. Our thanks to the many colleagues who have made contributions to this list of corrections. We fear, however, that other errors may still have escaped detection. *Caveat lector*!

Page *vi*, lines 16 and 18.   Löewner's   *should read*   Löwner's.

Page 13, Lemma 1-1. $K(\rho) \leq 1$ *should read* $K(\rho) \leq -1$.

Page 19, equation 1-29. $\dfrac{4 - \log |[f(z)]|}{4 - \log |[f(0)]|}$ *should read* $\dfrac{4 - \log |\zeta(f(z))|}{4 - \log |\zeta(f(0))|}$.

Page 21, line 6. $[58, \dots]$ *should read* $[59, \dots]$.

Page 29, line -4.   unit disk   *should read*   exterior of the unit disk.

Page 29, line -3.   comprises   *should read*   contains.

Page 29, line -2. $1 \geq \frac{1}{4}|b|$ *should read* $1 \geq 1/(4|b|)$.

Page 30, line 4. $\displaystyle\iint_\Omega \left[\left(\frac{\partial u}{\partial x}\right)^2 + \left(\frac{\partial u}{\partial y}\right)^2 \right] dxdy$ *should read*

$$\iint_\Omega \left[\left(\frac{\partial u}{\partial x}\right)^2 + \left(\frac{\partial u}{\partial y}\right)^2 \right] dxdy \,.$$

Page 32, line 1.   nonincreasing   *should read*   nondecreasing.

Page 36, lines 9, 10.   [59] and [54]   *should read*   [61] and [52], respectively.

Page 40, last line. $\{|z| > R\}$ *should read* $\{|z| > R\,,\ \text{Im } z > 0\}$.

Page 42, line 1.   [61]   *should read*   [6].

Page 43, line 14.

$$\frac{d \log M(r)}{d \log r} \leq \frac{4}{\pi \theta(r)} \log M(r) \quad \textit{should read} \quad \frac{d \log M(r)}{d \log r} \geq \frac{4}{\pi \theta(r)} \log M(r) \,.$$

Page 43, line 18. $\log[1 - \omega(r_0)] \leq \cdots$ *should read* $1 - \omega(r_0) \leq \cdots$.

Page 44, line -3. $\log M(r) \leq \dfrac{\log R}{\log r} \log M$ *should read* $\log M(r) \leq \dfrac{\log r}{\log R} \log M$.

159

Page 45, line 8.   [60]   *should read*   [62] .

Page 45, Lemma 3-1.   its minimum on $|z| = R$ at R, and its maximum on ...
  should read   its maximum on $|z| = R$ at R, and its minimum on ... .

Page 46, line 4. $g(\frac{R^m}{M})$ *should read* $g(-\rho, \frac{R^m}{M})$.

Page 46, equation after (3-9).

$$\frac{1}{2\pi} \int_C \cdots - \sum_1^N \omega(a_i) =   \text{ should read }   \frac{1}{2\pi} \int_C \cdots + \sum_1^N \omega(a_i) = .$$

Page 49, line 1. $|f(0)| \leq \cdots$ *should read* $|f(z_0)| \leq \cdots$.

Page 51, line 13. $L(\Gamma, \rho) = L(\Gamma', \rho')$ *should read* $L(\gamma, \rho) = L(\gamma', \rho')$.

Page 51, Definition 4-1.

$$\frac{\sup_\rho L(\Gamma, \rho)^2}{A(\Omega, \rho)}   \text{ should read }   \sup_\rho \frac{L(\Gamma, \rho)^2}{A(\Omega, \rho)} .$$

Page 67, lines 5,6. $d_\Omega(E_1, E_2) \geq 1/D(u)$ *should read* $d_\Omega(E_1, E_2) \leq 1/D(u)$.

Page 72, line 11. $e^{2\pi M(r)}$ *should read* $e^{2\pi M(R)}$.

Page 74, line -2. p *should read* $\wp$.

Page 77, line 9. E_1' and E_2' ...   *should read* $\widehat{E}_1'$ and $\widehat{E}_2'$ ... .

Page 78, Corollary. $\int_a^b dx/\theta(x) \geq \frac{1}{2}$ *should read* $\int_a^b dx/\theta(x) \geq 1$.

Page 79, line 16. $d(C_r, E) \leq \alpha$ *should read* $d(C_r, E) \leq \alpha/2\pi$.

Page 79, line -12.   (4-26)   *should read*   (4-24) .

Page 80, line -6.   Theorem 4-8   *should read*   Theorem 4-9 .

Page 80, line -4. $-(1/\pi) \log E$ *should read* $-(1/\pi) \log \operatorname{cap} E$.

Page 81, line -3.   Theorems 4-5 and 4-6   *should read*   Theorems 4-4 and 4-5 .

Page 84, line 1. $g(z) = zh(z^2)^{\frac{1}{2}}$ *should read* $g(z) = zh(z^2)$.

Page 88, inequality (5-11). $|c_1|^3$ *should read* $|c_1|^2$.

Page 88, displayed inequality between (5-12) and (5-13).   In the first term on the
  right-hand side, the denominator $3(1 - |b_1|^2)^2$ *should read* $3(1 - |b_1|^2)$.

Page 90, line 11.   Pommerenke [52,53]   *should read*   Pommerenke [53,54] .

Page 101, line 7. $\overline{\Gamma}(z_0, z) \frac{\rho^2 e^{i\alpha}}{t - z_0}$ *should read* $\overline{\Gamma}(z_0, z) \frac{\rho^2}{t - z_0}$.

Page 103, line -6. $e^{-i\alpha} \left[B(\zeta) + \frac{A(0)}{2} \right]$ *should read* $e^{-i\alpha} \left[B(\zeta) + \frac{\overline{A(0)}}{2} \right]$.

Page 110, line 15. $Q(z) = 0$ *should read* $Q(z) \geq 0$.

Page 146, equation (10-12). $S_r u_\rho - \operatorname{Re} \frac{1}{z}$ *should read* $S_r(u_\rho - \operatorname{Re} \frac{1}{z})$.

Page 146, line -3.   (4-13) and (4-14)   *should read*   (10-13) and (10-14) .

Page 153, reference [15].   Minimal" Surfaces   *should read*   Minimal Surfaces" .

ISBN 978-0-8218-5270-5

CHEL/371.H

About this book

Most conformal invariants can be described in terms of extremal properties. Conformal invariants and extremal problems are therefore intimately linked and form together the central theme of this classic book which is primarily intended for students with approximately a year's background in complex variable theory. The book emphasizes the geometric approach as well as classical and semi-classical results which Lars Ahlfors felt every student of complex analysis should know before embarking on independent research.

At the time of the book's original appearance, much of this material had never appeared in book form, particularly the discussion of the theory of extremal length. Schiffer's variational method also receives special attention, and a proof of $|a_4| \leq 4$ is included which was new at the time of publication. The last two chapters give an introduction to Riemann surfaces, with topological and analytical background supplied to support a proof of the uniformization theorem.

Included in this new reprint is a Foreword by Peter Duren, F. W. Gehring, and Brad Osgood, as well as an extensive errata.

Conformal invariants: topics in geometric function theory *encompasses a wealth of material in a mere one hundred and fifty-one pages. Its purpose is to present an exposition of selected topics in the geometric theory of functions of one complex variable, which in the author's opinion should be known by all prospective workers in complex analysis. From a methodological point of view the approach of the book is dominated by the notion of conformal invariant and concomitantly by extremal considerations. ...It is a splendid offering.*

—**Reviewed for** *Math Reviews* **by M. H. Heins in 1975**